Certificate Course

UNDERSTANDING MATHEMATICS 4

C. J. Cox & D. Bell

John Murray

© C. J. Cox and D. Bell 1986

First published 1986
by John Murray (Publishers) Ltd
50 Albemarle Street, London W1X 4BD

Reprinted 1987

Printed in Great Britain by
The Alden Press Ltd, Oxford

British Library Cataloguing in Publication Data

Cox, Christopher J.
  Understanding mathematics 4:
  certificate course.
  1. Mathematics—1961-
  I. Title   II. Bell, D. (David)
  510   QA39.2

ISBN 0-7195-4302-9

Foreword

Children's Understanding of Mathematics 11–16, which was published in 1981, reported the work of the mathematics team of the research project *Concepts in Secondary Mathematics and Science* (CSMS). Christopher Cox was head of department in one of the schools which co-operated with us in the research.

From the beginning he and his colleagues utilised the results we were able to make available. In particular, he noted the lack of success experienced by many children and the errors they committed. This led to the writing of material for use by the mathematics staff in his own and other schools, work which has led to this series of books.

The philosophy behind *Understanding Mathematics* is that, traditionally, the mathematics in secondary school becomes too difficult, too soon, for many children. Thus the exercises here provide a wide range of examples in order to cater for different learning rates, so organised that the learner is encouraged to take some responsibility for the amount of practice needed. The existence and benefits of the calculator and computer are recognised and exercises are provided for their use. Above all, the exercises are full of *interesting* questions and not just lists of sums. Also, the accompanying Teachers' Manuals contain a wide range of very useful copyable resource material, which is additional to the detailed guides to all the exercises.

CSMS showed that many children could not cope with the secondary mathematics they were expected to learn. Other research, particularly that reported in the publication of *Children Reading Mathematics* (edited by H. Shuard and A. Rothery) has pin-pointed the special problems faced in conveying meaning through written text. The writers of this series of books have taken the messages to heart and done something about it. Thus the reader is receiving the benefit of tested and well thought-out material written by enthusiastic and dedicated practising teachers.

Kath Hart

Preface

Understanding Mathematics is a complete course for secondary pupils in the 11–16 age range. It has proved to be equally effective with average pupils as with those of the highest ability, its exercise structure providing the flexibility needed to cater for this wide range. All pupils use the same books and follow the same common core, allowing easy transfer between sets and not labelling any pupil as 'lower ability'.

The two-book *Certificate Course* is designed to meet the needs of students preparing for GCSE mathematics. The content and achievement levels closely match the GCSE Mathematics Criteria.

Book 4 offers complete cover of level 2 (middle level) requirements. To ensure that the Certificate Course is self-contained, it incorporates comprehensive revision and self-assessment of previous years' work.

The Topic Matrix on pages x to xv of the Teachers' Manual shows the full plan of the Foundation and Certificate Courses.

The development of each topic was planned with reference to the findings of CSMS*, resulting in a common core with a less steep incline of difficulty than other texts, although the latter part of each exercise will challenge the highest ability pupil. Each topic is revised at each appearance before being developed further.

*Research project reported in K. Hart's *Children's Understanding of Mathematics 11–16*, John Murray

Acknowledgements

The authors are grateful for advice received from the following: Dr Kath Hart, Brian Bolt (Exeter University), Andrew Rothery (Worcester College of Education), Alec Penfold, Martyn Dunford (Huish Episcopi School), Jacqueline Gilday (Wells Blue School), Hazel Bevan (Millfield School), John Wishlade (Uffculme School) John Halsall, David Symes, Simon Goodenough, Mary Mears.

They are also indebted to Mr. R. A. Batts of Texas Instruments for checking the facts given in Chapter 4.

The authors thank their publishers for their help, their wives for their tolerance, and all the many teachers and pupils who have helped in the testing and revising of the course.

Illustrations by Tony Langham.

Contents

About this Book

This is the first of our two-book Certificate Course in mathematics for GCSE. All the work in it has been successfully tried by many different students.

Each chapter is concerned with a **topic** and is divided into **exercises**. Most chapters begin with a **You need to know** section which summarises the work you need to have covered, followed by **Test yourself** questions to check you know what you should, before setting out on the exercises.

The exercises have four kinds of question:

Introductory questions (Common Core) are for everyone.

Starred questions (Reinforcement) are optional for those who find the introductory questions very easy.

Further questions (Development) follow. These continue the topic to a higher level.

Boxed questions (Extension) challenge those who are keen and quick, and give lots of ideas for investigations and practical work.

This structure helps you to learn at your own pace and builds up your confidence.

To help you with homework and revision we have also provided a **Glossary** where you can check on words that you do not understand, **Answers** to the Test yourself and most other questions, and an **Index**.

Calculators will help you with many of the exercises.

Computer programs are included and the BASIC used will work on all the popular micros. The few changes needed for some machines are noted in the programs. We will be pleased to hear of any mathematical programs that you write and of any improvements that you make in ours.

1 Matrices: transformations

● You need to know . . .

● Co-ordinates, and the lines $y = 0$, $x = 0$, $y = x$ and $y = -x$

Point A has co-ordinates $(2, 0)$.

Point B has co-ordinates $(0, 1)$.

Point C has co-ordinates $(-1, -1)$.

Line DE is the x-axis. Its equation is $y = 0$.

Line FG is the y-axis. Its equation is $x = 0$.

The x-axis crosses the y-axis at the origin.

Line HI has the equation $y = x$. Each point on $y = x$ has its x-co-ordinate the same as its y-co-ordinate, e.g. $(3, 3)$; $(-2, -2)$; $(1.5, 1.5)$.

Line JK has the equation $y = -x$. Each point on $y = -x$ has its y-co-ordinate equal to its x-co-ordinate times -1, e.g. $(3, -3)$; $(0, 0)$; $(-2, 2)$.

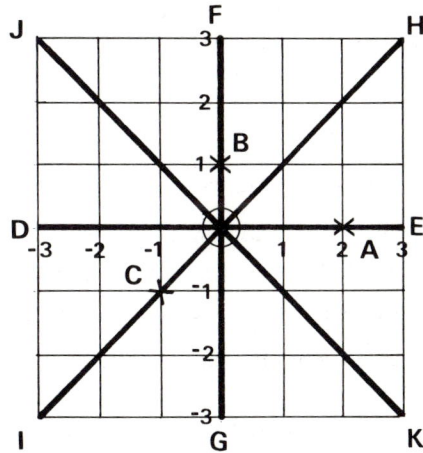

Fig. 1:1

● Multiplication of matrices

Constant times matrix

Example $3 \begin{pmatrix} 2 & -1 \\ 0 & 3 \end{pmatrix} = \begin{pmatrix} 6 & -3 \\ 0 & 9 \end{pmatrix}$

Matrix times matrix

Example $\begin{pmatrix} 2 & 3 \\ 2 & 5 \end{pmatrix} \begin{pmatrix} 4 & 2 \\ 1 & 5 \end{pmatrix} = \begin{pmatrix} 11 & 19 \\ 13 & 29 \end{pmatrix}$

Step One $(2 \quad 3) \begin{pmatrix} 4 \\ 1 \end{pmatrix} \rightarrow 8 + 3 = 11$

Step Two $(2 \quad 3) \begin{pmatrix} 2 \\ 5 \end{pmatrix} \rightarrow 4 + 15 = 19$

Step Three $(2 \quad 5) \begin{pmatrix} 4 \\ 1 \end{pmatrix} \rightarrow 8 + 5 = 13$

Step Four $(2 \quad 5) \begin{pmatrix} 2 \\ 5 \end{pmatrix} \rightarrow 4 + 25 = 29$

Be especially careful when negative (minus) numbers are involved. Check the following example carefully:

$\begin{pmatrix} 2 & -1 \\ -3 & 1 \end{pmatrix} \begin{pmatrix} -1 & -2 \\ 1 & 4 \end{pmatrix} = \begin{pmatrix} -3 & -8 \\ 4 & 10 \end{pmatrix}$

Remember that most matrices give different answers when multiplied together, depending on which one is at the front (we say multiplication of matrices is 'not commutative').

For example,

$$\begin{pmatrix} 1 & 0 \\ 1 & 2 \end{pmatrix}\begin{pmatrix} 2 & 1 \\ 0 & 0 \end{pmatrix} = \begin{pmatrix} 2 & 1 \\ 2 & 1 \end{pmatrix}$$

but

$$\begin{pmatrix} 2 & 1 \\ 0 & 0 \end{pmatrix}\begin{pmatrix} 1 & 0 \\ 1 & 2 \end{pmatrix} = \begin{pmatrix} 3 & 2 \\ 0 & 0 \end{pmatrix}$$

● The transformations of rotation, reflection, enlargement, shear and stretch

Rotation

To describe a rotation state the centre of rotation and the angle turned (+ve rotations are anticlockwise).

Example In Figure 1:2 triangle ABC has rotated through 270° about O.

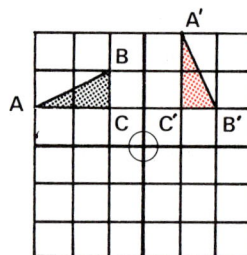

Fig. 1:2

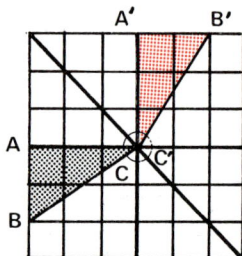

Fig. 1:3

Reflection

To describe a reflection state the equation, or the name, of the mirror line.

Example In Figure 1:3 triangle ABC has been reflected in the line $y = -x$.

Enlargement

To describe an enlargement state the scale factor by which the figure has been enlarged and the centre of the enlargement.

Figures 1:4 to 1:7 illustrate four kinds of enlargement. In each diagram △ABC is mapped onto △A′B′C′.

Figure 1:4 is an enlargement of scale factor 2 from centre O.

Figure 1:5 is an enlargement of scale factor $\frac{1}{2}$ from centre O.
(Note that we refer to the transformation as an enlargement even when the image is smaller than the object.)

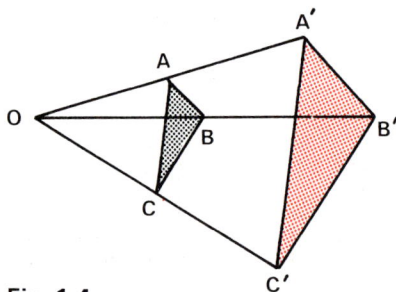

Fig. 1:4

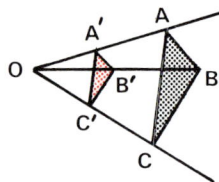

Fig. 1:5

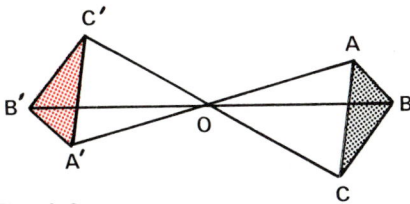

Fig. 1:6

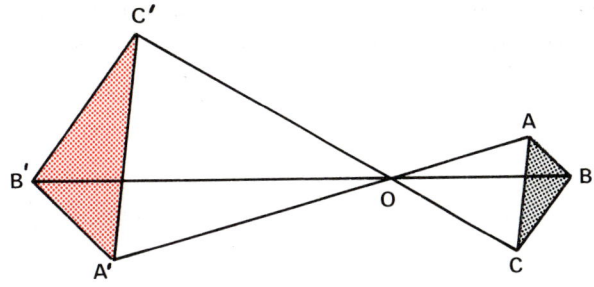

Fig. 1:7

Figure 1:6 is an enlargement of scale factor −1 from centre O.
(Note that in a negative enlargement the image is on the opposite side of the centre.)

Figure 1:7 is an enlargement of scale factor −2 from centre O.

Shear

The object is pushed sideways from a fixed line, which may be a side of the shape, or inside it, or outside it.

To describe a shearing state the equation, or the name, of the invariant (fixed) line and the original and transformed positions of one point.

Figures 1:8 to 1:10 show three shearings, each with line AB invariant.

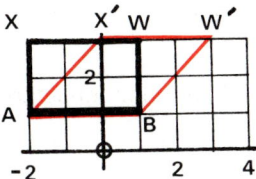

Fig. 1:8

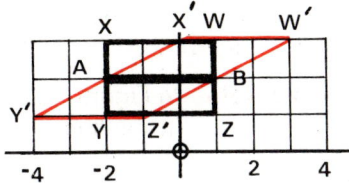

Fig. 1:9

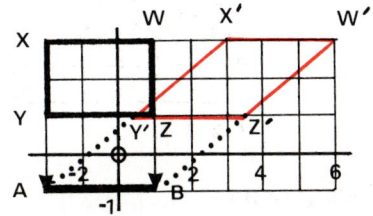

Fig. 1:10

Figure 1:8 is a shearing from AB, such that (1, 3) → (3, 3).

Figure 1:9 is a shearing from AB such that (1, 3) → (3, 3).

Figure 1:10 is a shearing from AB such that (1, 3) → (6, 3).

Stretch

The object is pulled to make it longer, or wider, or both, from a fixed line which may be a side, or inside, or outside the object.

To describe a stretch state the invariant line(s) and the stretch factor(s).

Figures 1:11 to 1:13 all show a stretch from AB of stretch factor 2.

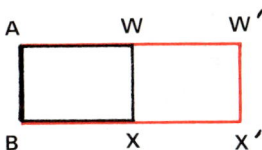

Fig. 1:11

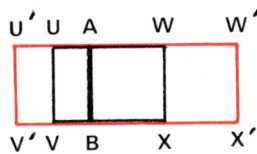

Fig. 1:12

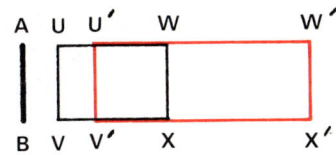

Fig. 1:13

Test yourself

1 State the co-ordinates of points A, B, C and D in Figure 1:14.

2 In Figure 1:14 which point (A, B, C or D) lies on the line:
(a) $y = 0$ (b) $x = 0$ (c) $y = x$ (d) $y = -x$?

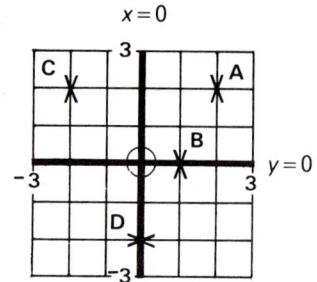

Fig. 1:14

3 Copy and complete the following co-ordinates so that the points will all lie on the line $y = x$.
(4,); (, $2\frac{1}{2}$); (, 0); (−3,); (, −1)

4 Repeat question 3, but this time make the points lie on the line $y = -x$.

5 Multiply out the following matrices:
(a) $(1 \quad 3) \begin{pmatrix} 2 \\ 4 \end{pmatrix}$ (b) $(2 \quad 1) \begin{pmatrix} 1 & 2 \\ 2 & 3 \end{pmatrix}$ (c) $\begin{pmatrix} 1 & 0 \\ 0 & 1 \end{pmatrix} \begin{pmatrix} 2 \\ 3 \end{pmatrix}$
(d) $\begin{pmatrix} 1 & -1 \\ 0 & 0 \end{pmatrix} \begin{pmatrix} 1 & -1 & 0 \\ 1 & 2 & -1 \end{pmatrix}$ (e) $\begin{pmatrix} -1 & 0 \\ 0 & -1 \end{pmatrix} \begin{pmatrix} -1 & 1 & 0 \\ 1 & -1 & -2 \end{pmatrix}$

6 $A = \begin{pmatrix} 1 & -1 \\ 0 & 1 \end{pmatrix}$ $B = \begin{pmatrix} 1 & 1 & 0 \\ 2 & 0 & 1 \end{pmatrix}$

(a) Find the product AB.
(b) Briefly explain why you cannot find the product BA.

7 In Figure 1:15 the hatched (///) square is transformed into the shaded square. Describe this transformation if it is:
(a) a rotation
(b) a reflection
(c) an enlargement
(d) a double reflection (give two possible answers).

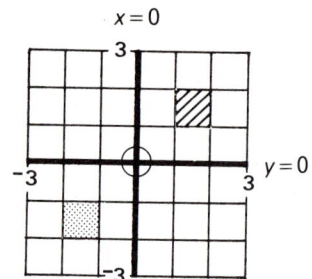

Fig. 1:15

8 Draw x- and y-axes from −4 to 4 each. Plot the triangle whose vertices are at (−2, 1), (−2, 3) and (−1, 3). Draw the image of this triangle after:
(a) a rotation of −90° about (0, 0)
(b) a reflection in $y = x$
(c) an enlargement of scale factor 2 from (0, 3)
(d) a shear with $x = -2$ (the vertical side) invariant, such that (−1, 3) → (−1, 4)
(e) a stretch from $y = 3$ (the horizontal side) of stretch factor 2.

9 The computer program "MATRIX2" (see Teachers' Manual for Book 3, page 27) can be used to practise matrix multiplication. You could develop the program so that it gives matrices for the operator to multiply, then have the computer check the answer.

10 Devise a computer program to illustrate one or more of the transformations revised in this chapter.

11 A computer can easily produce maps to different scales (within the limits of the original measurements). Many Ordnance Survey Maps are now produced by a computer. Devise a program to draw a plan of a building 15 metres long and 12 metres wide to any requested scale.

A Rotation and reflection

For Discussion

Example One

We can use the following three methods to define the position of points A, B and C on the grid in Figure 1:16.

Co-ordinates

A: (1, 1)

B: (1, 2)

C: (3, 1)

Position vectors

$$\overrightarrow{OA} = \begin{pmatrix} 1 \\ 1 \end{pmatrix}$$

$$\overrightarrow{OB} = \begin{pmatrix} 1 \\ 2 \end{pmatrix}$$

$$\overrightarrow{OC} = \begin{pmatrix} 3 \\ 1 \end{pmatrix}$$

Matrix

$$\begin{array}{c} \\ x \\ y \end{array} \begin{array}{ccc} A & B & C \\ \begin{pmatrix} 1 & 1 & 3 \\ 1 & 2 & 1 \end{pmatrix} \end{array}$$

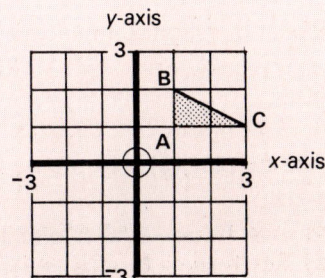

Fig. 1:16

Example Two

Using the matrix method to define the triangle, we can find the co-ordinates of the triangle after a rotation of 270° about (0, 0) by multiplying by the matrix $\begin{pmatrix} 0 & 1 \\ -1 & 0 \end{pmatrix}$:

$$\begin{pmatrix} 0 & 1 \\ -1 & 0 \end{pmatrix} \begin{array}{ccc} A & B & C \\ \begin{pmatrix} 1 & 1 & 3 \\ 1 & 2 & 1 \end{pmatrix} \end{array} = \begin{array}{ccc} A' & B' & C' \\ \begin{pmatrix} 1 & 2 & 1 \\ -1 & -1 & -3 \end{pmatrix} \end{array}$$

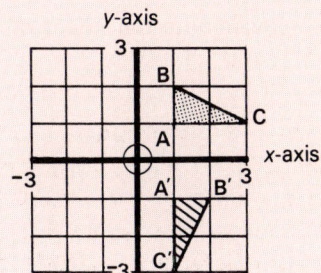

Fig. 1:17

1 Draw eight pairs of axes from -4 to 4 each. On each grid plot the quadrilateral whose vertices are at (1, 1), (4, 1), (4, 2) and (2, 2).

Plot the result of multiplying the matrix of the quadrilateral by each of the eight matrices $\begin{pmatrix} a & 0 \\ 0 & b \end{pmatrix} \begin{pmatrix} 0 & a \\ b & 0 \end{pmatrix}$ where $a, b \in \{1, -1\}$.

For example, if $a = -1$ and $b = 1$ the matrices are $\begin{pmatrix} -1 & 0 \\ 0 & 1 \end{pmatrix}$ and $\begin{pmatrix} 0 & -1 \\ 1 & 0 \end{pmatrix}$.

Describe each transformation fully.

*2 On one set of axes from -5 to 5 each, plot the triangle whose matrix is $\begin{pmatrix} 0 & 0 & 2 \\ 0 & 4 & 4 \end{pmatrix}$.

(a) Multiply this matrix by:

(i) $\begin{pmatrix} 0 & 1 \\ -1 & 0 \end{pmatrix}$ (ii) $\begin{pmatrix} -1 & 0 \\ 0 & -1 \end{pmatrix}$ (iii) $\begin{pmatrix} 0 & -1 \\ 1 & 0 \end{pmatrix}$.

(b) Plot the resulting transformations together on your grid. It will be clearer if you colour each image differently.

(c) Describe each transformation clearly, as an anticlockwise rotation.

*3 Repeat question 2, replacing the matrices in (a) by:

(iv) $\begin{pmatrix} -1 & 0 \\ 0 & 1 \end{pmatrix}$ (v) $\begin{pmatrix} 1 & 0 \\ 0 & -1 \end{pmatrix}$ (vi) $\begin{pmatrix} 0 & 1 \\ 1 & 0 \end{pmatrix}$ (vii) $\begin{pmatrix} 0 & -1 \\ -1 & 0 \end{pmatrix}$.

Note for part (c) that the transformations will not be rotations.

4 On axes from -6 to 6 plot the octagon with vertices, in order, at (0, 0), (6, 0), (6, -4), (4, -4), (4, -3), (5, -3), (5, -1) and (1, -1).

Plot, all on the same grid, the transformations caused by each of the matrices (i) to (vii) in questions 2 and 3.

5 Investigate the effect of multiplying $\begin{pmatrix} 0 & 1 & 0 \\ 0 & 0 & 1 \end{pmatrix}$ by any 2 by 2 matrix, e.g. $\begin{pmatrix} 3 & 4 \\ 5 & 6 \end{pmatrix}$.

6 Investigate the effect of multiplication by the following on the triangle represented by $\begin{pmatrix} 0 & 1 & 0 \\ 0 & 0 & 1 \end{pmatrix}$.

(a) $\begin{pmatrix} 1 & 0 \\ -1 & 1 \end{pmatrix}$ (b) $\begin{pmatrix} 1 & 0 \\ -2 & 1 \end{pmatrix}$ (c) $\begin{pmatrix} 1 & 3 \\ 0 & 1 \end{pmatrix}$ (d) $\begin{pmatrix} 1 & -1 \\ 0 & 1 \end{pmatrix}$ (e) $\begin{pmatrix} 1 & 2 \\ 0 & 1 \end{pmatrix}$

(f) $\begin{pmatrix} 1 & -2 \\ 0 & 1 \end{pmatrix}$ (g) $\begin{pmatrix} -2 & 0 \\ 0 & -2 \end{pmatrix}$ (h) $\begin{pmatrix} 2 & 0 \\ 0 & 1 \end{pmatrix}$ (i) $\begin{pmatrix} 1 & 0 \\ 0 & 2 \end{pmatrix}$

7 This computer program can be used to calculate the effect of a transforming matrix on a grid point.

```
  5  REM "TRANSMAT"
 10  DIM M(4)
 20  PRINT "Type in 2 by 2 transforming matrix."
 30  PRINT "First row, then second row."
 40  PRINT "Use RETURN after each of the four numbers." (or ENTER )
 50  FOR N = 1 TO 4
 60  INPUT M(N)
 70  NEXT N
 80  CLS (Clear screen)
 90  LET A = 1
100  PRINT "("; TAB(4 − LEN(STR$(M(A)))); M(A); TAB(9 − LEN(STR$(M(A +
     1)))); M(A + 1); ")"
110  LET A = A + 2
120  IF A = 3 THEN GOTO 100
130  PRINT
140  PRINT "x co-ordinate of object point?"
150  INPUT X
160  PRINT "y co-ordinate of object point?"
170  INPUT Y
180  PRINT "Image of ("; X; ","; Y; ") is at ("; M(1) ∗ X + M(2) ∗ Y; ","; M(3) ∗ X
     + M(4) ∗ Y; ")"
190  PRINT "More?? (Y/N)"
200  INPUT Y$
210  IF Y$ = "Y" THEN GOTO 80
220  GOTO 20
```

B   Enlargement; shear; stretch

For Discussion

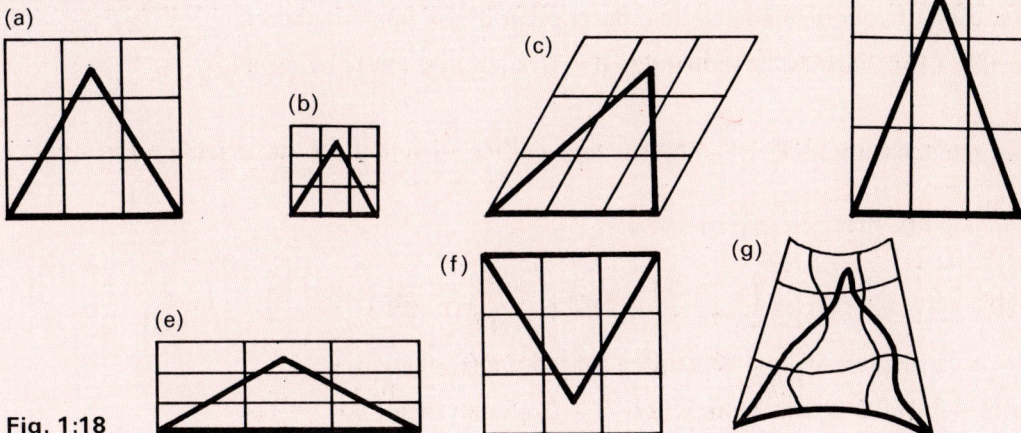

(a) (b) (c) (d) (e) (f) (g)

Fig. 1:18

1 (a) On axes from -4 to 6 plot the quadrilateral whose co-ordinate matrix is $\begin{pmatrix} 0 & 2 & 2 & 1 \\ 0 & 0 & 1 & 1 \end{pmatrix}$.

Repeat this on three more grids.

(b) Calculate $\begin{pmatrix} 2 & 0 \\ 0 & 2 \end{pmatrix}\begin{pmatrix} 0 & 2 & 2 & 1 \\ 0 & 0 & 1 & 1 \end{pmatrix}$ and plot the quadrilateral given by your answer on one of your grids.

(c) On your other three grids plot the transformation given by:

(i) $\begin{pmatrix} 3 & 0 \\ 0 & 3 \end{pmatrix}$ (ii) $\begin{pmatrix} -1 & 0 \\ 0 & -1 \end{pmatrix}$ (iii) $\begin{pmatrix} -2 & 0 \\ 0 & -2 \end{pmatrix}$.

2 On four grids, each from -3 to 6, plot the triangle with vertices at $(3, 0)$, $(6, 0)$ and $(6, 6)$.

Plot the transformations caused by:

(a) $\begin{pmatrix} \frac{1}{2} & 0 \\ 0 & \frac{1}{2} \end{pmatrix}$ (b) $\begin{pmatrix} \frac{1}{3} & 0 \\ 0 & \frac{1}{3} \end{pmatrix}$ (c) $\begin{pmatrix} -\frac{1}{2} & 0 \\ 0 & -\frac{1}{2} \end{pmatrix}$ (d) $\begin{pmatrix} -\frac{1}{3} & 0 \\ 0 & -\frac{1}{3} \end{pmatrix}$.

3 Question 1(b) shows *an enlargement, scale factor 2, centre (0, 0)*.
Question 2(c) shows *a negative enlargement, scale factor $-\frac{1}{2}$, centre (0, 0)*.

Label these two diagrams as given in italics above, then label the other diagrams in a similar way.

4 On six grids, x from -2 to 4, y from -3 to 1, plot the square with vertices at $(0, 0)$, $(1, 0)$, $(1, 1)$ and $(0, 1)$. (This is usually called **the unit square**.)

Investigate the transformation caused by:

(a) $\begin{pmatrix} 1 & 2 \\ 0 & 1 \end{pmatrix}$ (b) $\begin{pmatrix} 1 & 0 \\ -3 & 1 \end{pmatrix}$ (c) $\begin{pmatrix} 1 & -2 \\ 0 & 1 \end{pmatrix}$ (d) $\begin{pmatrix} 1 & 3 \\ 0 & 1 \end{pmatrix}$

(e) $\begin{pmatrix} 1 & 0 \\ -2 & 1 \end{pmatrix}$ (f) $\begin{pmatrix} 1 & -1 \\ 0 & 1 \end{pmatrix}$.

Under each of your diagrams write a description of the transformation.

Example (a) Shearing, invariant line $y = 0$, such that $(1, 1) \rightarrow (3, 1)$.

5 On six grids, each from 0 to 6, plot the square $\begin{pmatrix} 0 & 2 & 2 & 0 \\ 0 & 0 & 2 & 2 \end{pmatrix}$.

Investigate the effect on this square of:

(a) $\begin{pmatrix} 2 & 0 \\ 0 & 1 \end{pmatrix}$ (b) $\begin{pmatrix} 3 & 0 \\ 0 & 1 \end{pmatrix}$ (c) $\begin{pmatrix} 1 & 0 \\ 0 & 2 \end{pmatrix}$ (d) $\begin{pmatrix} 1 & 0 \\ 0 & 3 \end{pmatrix}$ (e) $\begin{pmatrix} 2 & 0 \\ 0 & 3 \end{pmatrix}$ (f) $\begin{pmatrix} 3 & 0 \\ 0 & 2 \end{pmatrix}$.

Under each diagram write a description of the transformation.

Examples (a) One-way stretch from $x = 0$, stretch factor 2.
(e) Two-way stretch from $x = 0$, stretch factor 2, and from $y = 0$, stretch factor 3.

6 Investigate the effects of other 2 by 2 matrices on the square in question 5.

7 Investigate the effect of two matrix multiplications, using the special matrices in Exercise 1A, question 1.

Example
$$\begin{pmatrix} 0 & -1 \\ 1 & 0 \end{pmatrix}\begin{pmatrix} 1 & 0 \\ 0 & 1 \end{pmatrix}\begin{pmatrix} 0 & 2 & 1 \\ 0 & 0 & 2 \end{pmatrix} = \begin{pmatrix} 0 & -1 \\ 1 & 0 \end{pmatrix}\begin{pmatrix} 0 & -2 & -1 \\ 0 & 0 & 2 \end{pmatrix}$$

$$= \begin{pmatrix} 0 & 0 & -2 \\ 0 & -2 & -1 \end{pmatrix}$$

2 Equations: using brackets

● You need to know . . .

● Basic algebraic notation

When using letters to stand for numbers we do not need to write multiplication signs.

Example $3 \times a \times b$ is usually written as $3ab$.

Note that $3 \times 4 \times 5$ can not be written as 345, and that a computer needs $3 * 4 * 5$ and $3 * a * b$.

When using letters to stand for numbers the division sign is usually replaced by writing the expression as a fraction.

Example $3 \div a$ is usually written as $\dfrac{3}{a}$.

Note that $3 \div 4$ can also be written as $\dfrac{3}{4}$ or 3/4, and that a computer needs 3/4 for both $3 \div 4$ and $\frac{3}{4}$.

● Directed numbers

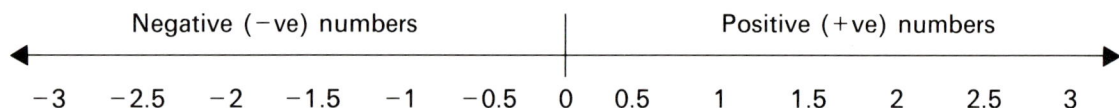

Negative (−ve) numbers ←———————————|———————————→ Positive (+ve) numbers

$$-3 \quad -2.5 \quad -2 \quad -1.5 \quad -1 \quad -0.5 \quad 0 \quad 0.5 \quad 1 \quad 1.5 \quad 2 \quad 2.5 \quad 3$$

Positive (plus) numbers need no signs; negative (minus) numbers need − signs.

Like signs multiply to make a plus:

$$- \; -3 \to +3 \qquad -3 \times -2 \to +6$$

Unlike signs multiply to make a minus:

$$- \; +3 \to -3 \qquad + \; -3 \to -3 \qquad -3 \times +2 \to -6 \qquad +3 \times -2 \to -6$$

Examples   If $a = -2$, $b = -3$, and $c = 4$, then:

$$a + b = -2 + -3 \to -2 - 3 = -5$$

> **Note:** $-2 - 3$ can be thought of as 'down 2 then down another 3'. The two minuses do not make a plus here, as they are not multiplied.

$$bc = -3 \times 4 = -12$$

$$ab = -2 \times -3 = 6$$

$$\frac{c}{a} = 4 \div -2 = -2 \qquad \textbf{Note: } \text{The same rules apply for division as for multiplication.}$$

$$\frac{a}{b} = -2 \div -3 = \tfrac{2}{3}$$

● Brackets

Any term written directly before a bracket multiplies each term in the bracket.

Examples $2(4 + a) \rightarrow 8 + 2a$ Working: $2 \times 4 = 8$; $2 \times a = 2a$

Note: Read the $\rightarrow$ sign in this course as 'becomes'.

$-2(4 + a) \rightarrow -8 - 2a$ Working: $-2 \times 4 = -8$; $-2 \times a = -2a$

$-2(4 - a) \rightarrow -8 + 2a$ Working: $-2 \times 4 = -8$; $-2 \times -a = +2a$

$-(4 + a) \rightarrow -4 - a$ Working: $-(4 + a) \Rightarrow -1(4 + a)$; $-1 \times 4 = -4$; $-1 \times a = -a$

Note: $\Rightarrow$ is the sign for 'implies' or 'means'.

$a + 2(b - c) \rightarrow a + 2b - 2c$

$a - 2(b - c) \rightarrow a - 2b + 2c$

Note that a computer requires the $*$ (multiply) sign between the term and the bracket, e.g. $-2 * (4 - a)$.

⬤ Inspection method for equation solving

Equations with only one letter-term are nearly always solved (to find the value of the letter) most easily by the 'inspection' approach, not by using 'rules'.

Examples If $b - 2 = 8$ then b must be $\underline{10}$ (as $10 - 2 = 8$).

If $8 + c = 6$ then c must be $\underline{-2}$ (as $8 + -2 = 6$).

If $7 - 2x = 8$ then $2x$ must be -1 (as $7 - -1 \rightarrow 7 + 1 = 8$)
   so x must be $\underline{-\frac{1}{2}}$ (as $2 \times -\frac{1}{2} = -1$).

If $3e = 2$ then e must be $\frac{2}{3}$ (for 3 thirds make 1 whole, so 3 times two-thirds makes 2 wholes).

If $\dfrac{24}{1 - n} = 8$ then $1 - n$ must be 3 (as $24 \div 3 = 8$)
   so n must be $\underline{-2}$ (as $1 - -2 \rightarrow 1 + 2 = 3$)

If $4(2a - 7) = 12$ then $2a - 7$ must be 3 (as $4 \times 3 = 12$)
   so $2a$ must be 10 (as $10 - 7 = 3$)
   so a must be $\underline{5}$ (as $2 \times 5 = 10$).

The above solutions can be written as follows:

$b - 2 = 8 \Rightarrow \underline{b = 10}$

$8 + c = 6 \Rightarrow \underline{c = -2}$

$7 - 2x = 8 \Rightarrow 2x = -1 \Rightarrow \underline{x = -\frac{1}{2}}$

$3e = 2 \Rightarrow \underline{e = \frac{2}{3}}$

$\dfrac{24}{1 - n} = 8 \Rightarrow 1 - n = 3 \Rightarrow \underline{n = -2}$

$4(2a - 7) = 12 \Rightarrow 2a - 7 = 3 \Rightarrow 2a = 10 \Rightarrow \underline{a = 5}$

● Solution of simultaneous equations

When there are two unknown letters to be found you need two equations. For instance, $x + y = 8$ is true for an infinite number of pairs of values for x and y. But if we also know that $x = y + 2$ then the only possible solution is $x = 5$ and $y = 3$.

Four methods are possible:

One **Draw intersecting graphs.**

Two **Substitute for one letter its value in the other equation.** This is best used when one of the equations is in the form $x =$ or $y =$. For example, $y = x - 6$ and $3y - 2x = 8$.

Three **Add or subtract the equations to eliminate one of the letter-terms**, having multiplied as necessary to make one letter-term the same absolute value in both. ('Absolute' means 'ignoring the sign'; ABS on a BASIC computer.)

Four **Use matrices.** This is taught in Exercise 12B.

Method One **To solve simultaneously $y = 2x - 1$ and $y - x = 1$.**

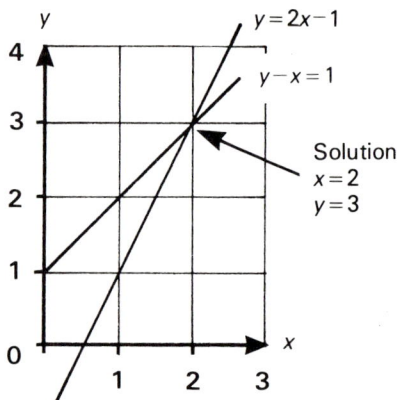

Fig. 2:1

Method Two **To solve $x = 2$ and $2x + 2y = 11$ simultaneously.**

We can think of this as finding where the line $x = 2$ crosses the line $2x + 2y = 11$.
Substitute the value $x = 2$ into $2x + 2y = 11$, giving $4 + 2y = 11$.
Then $4 + 2y = 11 \rightarrow 2y = 7 \rightarrow y = 3\frac{1}{2}$.
Answer: $\underline{\underline{x = 2, y = 3\frac{1}{2}}}$ (The lines cross at $(2, 3\frac{1}{2})$.)

To solve $y = x - 6$ and $x + y = 4$ simultaneously.

Substitute the value $y = x - 6$ into $x + y = 4$, giving $x + (x - 6) = 4$.
Then $x + (x - 6) = 4 \rightarrow x + x - 6 = 4 \rightarrow 2x - 6 = 4 \rightarrow 2x = 10 \rightarrow x = 5$.
We know that $y = x - 6$, so if $x = 5$ then y must be -1.
Answer: $\underline{\underline{x = 5, y = -1}}$.

To solve **$x = 2y - 1$ and $y + 2x = 2$ by substitution.**

Substitute for x, giving $y + 2(2y - 1) = 2$.
$y + 2(2y - 1) = 2 \rightarrow y + 4y - 2 = 2 \rightarrow 5y - 2 = 2 \rightarrow 5y = 4 \rightarrow y = \frac{4}{5}$.
Now as we know that $x = 2y - 1$, then $x = \frac{8}{5} - 1 = \frac{3}{5}$.
Answer: $\underline{x = \frac{3}{5}, \ y = \frac{4}{5}}$.

Method Three **To find the values of x and y that satisfy both $x + 3y = 9$ and $x - 2y = -1$.**

$$x + 3y = \ \ 9$$
$$\underline{x - 2y = -1} \quad \text{SUB}$$
$$\underline{\ \ \ 5y = 10}$$
$$\underline{\ \ \ y = 2}$$

Note: By subtracting, the x terms have been eliminated. If the two given equations are true, then the result of the subtraction is true.

Substitute $y = 2$
into $x + 3y = 9$,
giving $x + 6 = 9$,
so $\underline{x = 3}$.

Compare: $2 + 3 = 5$
$\underline{1 + 2 = 3} \quad \text{SUB}$
$\underline{1 + 1 = 2}$

Check both equations are true
for $x = 3$, $y = 2$:
$3 + 6 = 9$; $3 - 4 = -1$.

When subtracting,
$+3y - {-2y} \rightarrow +3y + 2y = 5y$
and $9 - {-1} \rightarrow 9 + 1 = 10$.

To solve $3x + 5y = 21$ and $7x - 2y = 8$ simultaneously.

$$3x + 5y = 21 \xrightarrow{\times 2} 6x + 10y = 42$$
$$7x - 2y = 8 \xrightarrow{\times 5} \underline{35x - 10y = 40} \quad \text{ADD}$$
$$\underline{41x \qquad \quad = 82}$$
$$\underline{x = \ \ 2}$$

Substitute $x = 2$ into $3x + 5y = 21 \rightarrow 6 + 5y = 21 \rightarrow 5y = 15 \rightarrow \underline{\underline{y = 3}}$.

Check this for yourself.

Test yourself

1 In this question, $m = -1$, $n = -2$, and $p = 3$. Evaluate (find the numerical value of):
(a) $3m$ (b) $p - n$ (c) $m - n$ (d) $m + p$ (e) $m + n$ (f) mp
(g) mn (h) $\dfrac{p}{m}$ (i) $\dfrac{m}{n}$.

2 Multiply out (as in the examples on page 11):
(a) $3(x + 4)$ (b) $2(a - 5)$ (c) $4(3 + 2a)$ (d) $2(2a - 5)$ (e) $-(3 + a)$
(f) $-(h - 1)$ (g) $-3(2x + 1)$ (h) $-2(x - 4)$ (i) $-3(3 + 4a)$.

3 Solve (find the value of the letter):
(a) $4z = -12$ (b) $12d = 3$ (c) $3r - 5 = 4$ (d) $6y - 6 = 0$ (e) $e + 6 = 4$
(f) $x + 3 = -4$ (g) $2t + 4 = -8$ (h) $4u + 1 = 6$ (i) $3 + 2m = -2$
(j) $17 - 3d = 1$ (k) $13 = 7 - 2b$ (l) $3 = 1 - 3v$.

2

*4 Multiply out:
(a) $3(2 + 3a)$ (b) $3(3a - 4)$ (c) $-(x + 1)$ (d) $-2(x + 4)$ (e) $-4(2 - a)$
(f) $-2(2a - 3)$ (g) $2(3a - 8)$.

*5 Solve:
(a) $4k - 9 = 7$ (b) $8 - 3w = 2$ (c) $24 + 3x = 3$ (d) $5t + 13 = 3$
(e) $4 - 2n = 8$ (f) $5 - 3q = -4$ (g) $4x - 18 = 16$ (h) $2 - 3r = 5$
(i) $3y + 12 = 5$ (j) $-11 = 6x + 7$ (k) $14 = -3a + 2$ (l) $-3 = -4f + 5$.

6 Solve:
(a) $\dfrac{x}{4} = 3$ (b) $\dfrac{d}{8} = -2$ (c) $\dfrac{15}{u} = 3$ (d) $\dfrac{16}{e} = -4$ (e) $\dfrac{18}{2t} = 9$.

7 Solve:
(a) $\dfrac{x}{4} + 6 = 14$ (b) $\dfrac{3s}{5} + 4 = 10$ (c) $16 - \dfrac{f}{3} = 12$ (d) $6 - \dfrac{2w}{3} = 4$
(e) $\dfrac{20}{n} - 4 = 6$ (f) $12 - \dfrac{25}{2x} = 7$.

8 Solve simultaneously by substitution:
(a) $x = y + 3$ and $x + 2y = 21$
(b) $y = 3x - 4$ and $x + y = 4$
(c) $y = 3x - 1$ and $x - 2y = -13$.

9 Solve simultaneously by elimination:
(a) $2x - 3y = 5$ and $x + 3y = -2$
(b) $3x - 2y = 10$ and $x - 2y = 2$
(c) $a + b = 15$ and $2a - 4b = -3$
(d) $3s - 2t = 5$ and $2s - 5t = 10$.

---

10 **Examples** $-2(4a - 5) + 3 \rightarrow -8a + 10 + 3 \rightarrow -8a + 13$

$3(a - 6) + 2(3 - a) \rightarrow 3a - 18 + 6 - 2a \rightarrow a - 12$

Simplify:
(a) $2(a + 3) + 3(2 - a)$ (b) $3(2 + 5a) - 4(2a - 3)$
(c) $5(a - 3) - 4(a + 5)$ (d) $2(a + 4) + 3(6 - 2a)$
(e) $-8(2 - 5a) - 4a - 9$ (f) $-5 - (3a - 7) + 2(1 - 8a) + a$.

11 Solve:
(a) $\dfrac{3x}{5} = -4$ (b) $\dfrac{12}{3w} = 8$ (c) $\dfrac{46}{7g} = 2$ (d) $3 = \dfrac{2}{3x} - 18$.

12 Solve simultaneously:
(a) $2a - \dfrac{b}{3} = 4$ and $3a + \dfrac{b}{2} = 12$ (b) $\dfrac{a}{10} + b = 2$ and $\dfrac{a}{2} - 2b = 3$
(c) $6a = 2b + 9$ and $3a + 4b = 12$
(d) $3a + 7b = 1$ and $5b + 2a + 1 = 0$.

Equations with two letter-terms

For Discussion

In solving equations like $3n - 2 = 5 + 2n$ we cannot use the inspection approach (see page 11) until one of the two letter-terms has been removed.

We remove one letter-term by adding to or subtracting from both sides of the equation a term which reduces it to zero, as in the following examples.

Examples (a) $3n - 2 = 5 + 2n \xrightarrow{-2n \text{ on both sides}} 3n - 2 - \mathbf{2n} = 5 + 2n - \mathbf{2n} \rightarrow n - 2 = 5$

(b) $5 - 2w = 2 + 3w \xrightarrow{+2w \text{ on both sides}} 5 - 2w + \mathbf{2w} = 2 + 3w + \mathbf{2w}$
$\rightarrow 5 = 2 + 5w$

(c) $4 - 3x = 5 - 2x \xrightarrow{+3x \text{ on both sides}} 4 - 3x + \mathbf{3x} = 5 - 2x + \mathbf{3x} \rightarrow 4 = 5 + x$

(d) $4 - 3z = 5 + 2z \xrightarrow{+3z \text{ on both sides}} 4 - 3z + \mathbf{3z} = 5 + 2z + \mathbf{3z} \rightarrow 4 = 5 + 5z$

It does not really matter which letter-term you reduce to zero, but the remainder of the solution is usually easier if you leave a positive letter-term, as we did in the above examples. This can be remembered by the following rule, if you like rules!

Remove the term with the smaller coefficient.

The 'coefficient' is the number in front of it. Note that in example (c), -3 is smaller than -2.

Note: Because we perform the same operation on each side of the equation it remains true or 'in balance'. We can see this in number statements:

$3 + 2 = 5 \xrightarrow{-2 \text{ on both sides}} 3 + 2 - \mathbf{2} = 5 - \mathbf{2} \rightarrow 3 = 5 - 2$

$3 - 2 = 1 \xrightarrow{+2 \text{ on both sides}} 3 - 2 + \mathbf{2} = 1 + \mathbf{2} \rightarrow 3 = 3$

Examples Solve $6n + 7 = 4n + 13$.
$6n + 7 = 4n + 13 \xrightarrow{-4n} 2n + 7 = 13 \Rightarrow 2n = 6 \Rightarrow \underline{n = 3}$

Solve $7 - 5x = 4x - 2$.
$7 - 5x = 4x - 2 \xrightarrow{+5x} 7 = 9x - 2 \Rightarrow 9x = 9 \Rightarrow \underline{x = 1}$

1 Solve the following equations.
 (a) $3y - 5 = 2y + 3$ (b) $6t - 4 = 5 + 3t$ (c) $4e + 8 = 11e + 15$
 (d) $12 - 4c = 8 - 2c$ (e) $6x + 10 = 13 - 12x$ (f) $6 - 5n = 4n + 24$

*2 Find the solution of:
 (a) $x + 8 = 3x + 4$ (b) $9 + 2k = 4k + 11$ (c) $3 - 2t = 4t + 15$
 (d) $4f + 1 = 10 - 2f$ (e) $3g + 2 = g + 2$ (f) $1 - 3h = 2\frac{1}{2} - 2h$.

3 **Example** Solve $3(x + 2) = 2(x - 1)$.
$3(x + 2) = 2(x - 1) \rightarrow 3x + 6 = 2x - 2 \xrightarrow{-2x} x + 6 = -2 \Rightarrow \underline{\underline{x = -8}}$.

Solve:
 (a) $2(x - 3) = x + 4$ (b) $2(x + 2) = x + 5$ (c) $3(3w - 4) = 4(w + 2)$.

4 Solve:
(a) $2(4a - 6) = 5(2a - 2)$ (b) $5(2 + 2k) = 2(3 + 3k)$ (c) $4(4r - 6) = 8(r + 3)$.

5 Solve:
(a) $4(3 + 2h) = 6(5 + h)$ (b) $3(2x - 4) = 5(-2x + 4)$.

6 If necessary, 'collect terms' on each side of the equation before beginning to solve it.

Example Solve $3(a - 2) - 1 = 4(a + 4) + 2a - 2$.

First multiply out the brackets and collect like terms:
$$3(a - 2) - 1 = 4(a + 4) + 2a - 2 \rightarrow 3a - 6 - 1 = 4a + 16 + 2a - 2$$
$$\rightarrow 3a - 7 = 6a + 14.$$

Now solve the equation:
$$3a - 7 = 6a + 14 \xrightarrow{-3a} -7 = 3a + 14 \Rightarrow 3a = -21 \Rightarrow \underline{\underline{a = -7}}$$

Solve:
(a) $1 + 3(n - 1) = 2n + 4$ (b) $7(s - 3) = 4(s + 2) - 2$
(c) $3e - 4(3e - 2) = 3(2 - 3e) + 2 - 2e$.

7 Solve:
(a) $4(r - 3) - 3(2 - r) = 2r + 7$ (b) $7(2p - 5) + 8(3p - 4) - 4(7p + 2) = 0$
(c) $9(6x + 7) - 4(2x + 5) = 40 - (x - 3)$
(d) $5y - 3(y - 2) = 8 - 2(2 - 5y) + 2(y - 5)$.

---

8 (a) Josie is 5 and her dad is 25. Using axes labelled 'Josie's age' (j-axis, from 0 to 50) and 'Dad's age' (d-axis, from 20 to 70), draw a straight line to represent the connection between their two ages.

(b) Draw the line $d = 2j$. Explain the significance of the intersection of this line with the line in part (a).

(c) Is dad ever three times Josie's age?

(d) Write and solve an equation to find when dad is $1\frac{1}{2}$ times Josie's age.

9 An aeroplane averaging 300 knots (300 nautical miles per hour) reaches its destination 10 min early, whilst if it averages 240 knots it is 6 min late. What is the planned time for the journey?

10 Isaac forgets to write the units, and makes the area of a square equal to its perimeter. What sizes could the square be?

11 An isosceles triangle has sides $x - 2$ cm; $3x - 7$ cm and $\frac{2}{3}x$ cm. What are the possible values of x? Draw each possible triangle.

12 A card box with no lid has to have a square base x cm by x cm, with upright sides y cm. What values of x and y will use the smallest area of card and give the box a volume of 4000 cm³?

3 Sets: three sets

● You need to know . . .

● Set notation and terminology

'Describe set A'. A = {the first four odd integers}

'Describe set B'. B = {triangular numbers less than 10}

'List set A'. A = {1, 3, 5, 7}

'List set B'. B = {1, 3, 6}

Symbols	Meaning
$3 \in A$	3 is an element (or member) of set A.
$2 \notin A$	2 is not an element of set A.
$n(A) = 4$	There are 4 elements in set A.
$\varnothing$ or { }	A null, or empty, set. It has no elements, e.g. {even numbers in set A}.
$\{1, 5\} \subset A$	{1, 5} is a subset of set A.
$A \supset \{5\}$	Set A contains the subset {5}.
$B \not\subset A$	Set B is not a subset of set A.
$A \cap B = \{1, 3\}$	The intersection of sets A and B. It consists of the elements common to both sets.
$A \cup B = \{1, 3, 5, 6, 7\}$	The union of sets A and B. It is made by joining the two sets together. Note that elements common to both sets are only written once in the union.
ξ	The universal set. It defines all the elements that may be used in a particular problem.
A'	The complement of set A. It contains those elements that are not in set A.

● Venn diagrams

Venn diagrams illustrate sets.

In Figure 3:1, A = {1, 3, 5, 7} and T = {1, 7}.
Set T is a subset of A ($T \subset A$ or $A \supset T$).

Fig. 3:1

In Figure 3:2, set A intersects set B.

$A \cap B = \{1, 3\}$ $n(A \cap B) = 2$

$A \cup B = \{1, 3, 5, 6, 7\}$

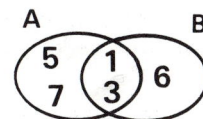

Fig. 3:2

In Figure 3:3, sets G and T are disjoint. They do not intersect.

$G \cap T = \varnothing$ $n(G \cap T) = 0$

Fig. 3:3

In Figure 3:4, ξ (the rectangle) is the universal set. The hatched region is D′, the complement of set D.

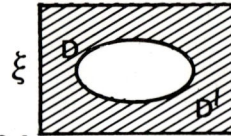

Fig. 3:4

Figure 3:5 shows the number of elements in sets ξ, A and B. It does not tell you what the elements are.

$n(A) = 2 + 2 = 4 \qquad n(A') = 1 + 3 = 4$

Note: $[n(A) + n(B)] - [n(A \cup B)] = n(A \cap B).$
$\qquad [\ 4\ +\ 3\ ] - [\ \ \ 5\ \ \ ] =\ \ \ \ 2$

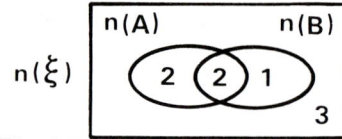

Fig. 3:5

Test yourself

1 ξ = {1, 2, 3, 4, 5, 6}
A = {2, 4, 6}   B = {1, 3, 5, 6}   D = {2, 4}

State whether each of the following statements is true or false.
(a) 4 ∈ A   (b) 2 ∉ D   (c) A ∩ B = ∅   (d) A ∪ D = A   (e) B′ = D
(f) D ⊄ A   (g) A ⊂ ξ   (h) B ⊅ D   (i) A ⊃ D   (j) D ⊄ ξ

2 Copy Figure 3:6 six times. Using one copy for each part, hatch (///):
(a) S ∪ T   (b) S ∩ T   (c) S′   (d) T′
(e) (S ∪ T)′   (f) (S ∩ T)′.

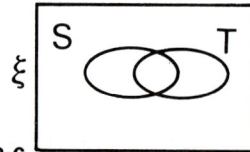

Fig. 3:6

3 Copy Figure 3:7 four times. Using one copy for each part, hatch:
(a) A ∩ B   (b) A ∪ B   (c) A′   (d) B′.

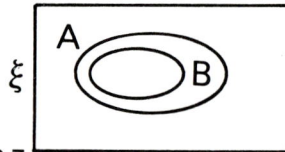

Fig. 3:7

4 For the elements in Figure 3:8, list
(a) P ∩ Q   (b) P ∪ Q   (c) P′   (d) Q′
(e) (P ∩ Q)′   (f) (P ∪ Q)′.

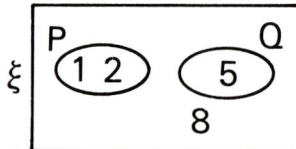

5 ξ = {2, 3, 4, 5}   M = {2, 5}   N = {3, 5}

Fig. 3:8

List:
(a) M ∩ N   (b) M ∪ N   (c) M′   (d) N′   (e) (M ∩ N)′   (f) (M ∪ N)′.

6 Draw Venn diagrams to show:
(a) A = {1, 2, 3, 4} and B = {2, 4, 6}
(b) P = {a, b, c, d} and Q = {a, b}
(c) M = {5, 6, 7} and N = {2, 3, 4}
(d) C = {cat, cow} and F = {cow, sheep}.

7 For Figure 3:9 state:
(a) $n(\xi)$ (b) $n(A)$ (c) $n(B)$
(d) $n(A \cap B)$ (e) $n(A \cup B)$ (f) $n(A')$
(g) $n(B')$ (h) $n(A \cap B)'$ (i) $n(A \cup B)'$.

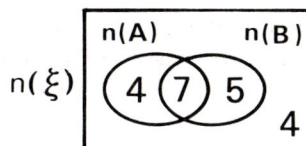

$n(\xi)$
n(A)   n(B)
4 (7) 5
4

Fig. 3:9

8 ξ = {class 4A}   H = {hockey players}   C = {cricket players}

State which of descriptions V to Z are correct for sets (a) to (e).

V = {neither hockey nor cricket players}
W = {not hockey players}
X = {hockey and/or cricket players}
Y = {not players of *both* hockey and cricket}
Z = {not cricket players}

(a) H′   (b) C′   (c) H ∪ C   (d) (H ∪ C)′   (e) (H ∩ C)′

---

9 P and Q are intersecting sets. Draw, hatch, and label separate Venn diagrams to show:
(a) $(P \cup Q)'$ (b) $(P \cap Q)'$ (c) $P' \cap Q$ (d) $P \cap Q'$ (e) $P \cup Q'$
(f) $P' \cup Q$ (g) $P' \cup Q'$ (h) $P' \cap Q'$.

10 Repeat question 9 with
(i) P and Q disjoint   (ii) P as a subset of Q.

A   Two-set problems

Example   Draw a Venn diagram to show: $n(A \cup B) = 16$, $n(A) = 9$, and $n(B) = 10$.

As $n(A) + n(B) = 19$, but $n(A \cup B) = 16$, then $n(A \cap B) = 19 - 16 = 3$. Fill in the 3, then the 6 and the 7, giving the answer shown in Figure 3:10.

n(A)   n(B)
6 (3) 7

Fig. 3:10

1 Draw a Venn diagram to show:
(a) $n(A \cup B) = 12$, $n(A) = 9$, $n(B) = 5$ (b) $n(C \cup D) = 11$, $n(C) = 10$, $n(D) = 7$
(c) $n(E \cup F) = 16$, $n(E) = 13$, $n(F) = 11$ (d) $n(G \cup H) = 14$, $n(G) = 6$, $n(H) = 12$.

2 Of 13 footballers, 8 wear Supaboots (B) and 9 wear shin-pads (P). All wear at least one of these. Illustrate this with a Venn diagram, then state how many wear:
(a) Supaboots and shin-pads   (b) Supaboots but not shin-pads.

3 Of 20 boats, 14 had sails (S) and 12 had engines (E). None had neither. Use a Venn diagram to find how many had:
(a) both engine and sails   (b) sails only.

3

4 In a holiday brochure, 21 hotels had a swimming-pool (P), 39 had a restaurant (R), and 12 had both swimming-pool and restaurant. How many had at least one of these two facilities?

5 In a class, 40% of the pupils had been to Germany, 30% to France, and 10% to both countries.
If there are 30 pupils, how many had:
(a) been only to Germany (b) been only to France
(c) been to both countries (d) not been to either country?

B Three-set Venn diagrams

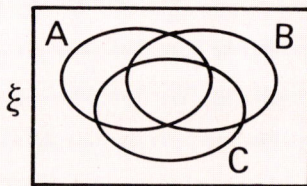

When there are three sets in a diagram, all possible intersections are allowed for if they are drawn as in Figure 3:11.

Fig. 3:11

1 Figure 3:12 illustrates the books liked by members of a form. List all the pupils who like:
(a) westerns (b) westerns and historical novels (c) westerns and science fiction
(d) none of these types of book (e) science fiction only
(f) westerns and science fiction, but not historical novels
(g) science fiction and historical novels, but not westerns.

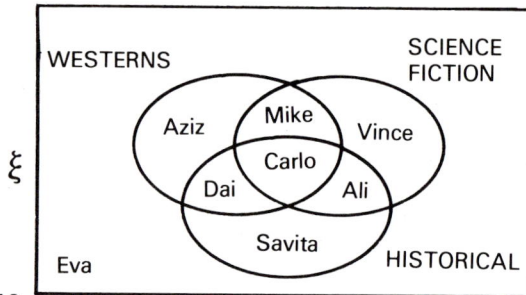

Fig. 3:12

2 Copy Figure 3:11 seven times and hatch:
(a) A ∩ B ∩ C (b) A ∩ B (c) B′ (d) A ∪ B (e) (A ∪ B)′
(f) (A ∩ C)′ (g) (A ∪ B ∪ C)′.

*3 List the following sets shown in Figure 3:13:
(a) ξ (b) A (c) B (d) A ∩ B ∩ C
(e) A ∩ B (f) A ∩ C (g) B′ (h) A ∪ B
(i) B ∪ C (j) (B ∪ C)′ (k) A ∪ B ∪ C.

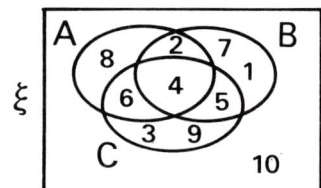

Fig. 3:13

20

4 (a) Copy Figure 3:11, then write in the elements so that:
$\xi = \{a, b, c, d, e, f, g, h, i, j\}$, $A = \{a, b, d, e, i\}$,
$B = \{b, c, e, f, h\}$, $C = \{d, e, f, g\}$.

(b) List:
(i) $A \cap B \cap C$ (ii) $A \cap B$ (iii) $A \cap C$ (iv) $B \cap C$ (v) $(A \cup B \cup C)'$.

5 (a) Copy Figure 3:11, then write in the elements so that:
$\xi = \{$integers from 1 to 15$\}$, $A = \{$prime numbers$\}$,
$B = \{1, 2, 3, 5, 8, 15\}$, $C = \{$multiples of 3$\}$.

(b) List:
(i) $A \cap B \cap C$ (ii) $A \cap C$ (iii) $B \cap C$ (iv) $A \cap (B \cup C)'$ (v) $B \cap (A \cup C)'$
(vi) $C \cap (B \cup A)'$ (vii) $(A \cup B \cup C)'$.

---

6 Hatch copies of Figure 3:11 to show:
(a) $A \cap (B \cup C)'$ (b) $A \cup (B \cap C)$ (c) $B \cup (A \cap C)$
(d) $C \cup (A \cap B)$ (e) $(A \cup B) \cap C'$ (f) $(A \cap C) \cap B'$.

7 Design a suitable Venn diagram to show all possible intersections for four sets.

Is there a limit to the number of sets for which it is possible to draw a 'universal' Venn diagram?

C Three-set problems

Venn diagrams can clarify complex logic problems. Follow the reasoning used in this example.

Maria has 32 dresses. Of these, 17 have sleeves, 22 have zips, and 15 have belts. 10 have zips and sleeves, 8 have zips and belts, 6 have sleeves and belts. All the dresses have at least one of these features.

Find how many have:
(a) only sleeves (b) only a zip (c) only a belt.

Summary using set notation
$n(\xi) = 32$; $n(S) = 17$; $n(Z) = 22$; $n(B) = 15$; $n(Z \cap S) = 10$; $n(Z \cap B) = 8$;
$n(S \cap B) = 6$; $n(S \cup B \cup Z)' = 0$.

Solution Method One
Insert the zero from $n(S \cup B \cup Z)' = 0$.
Let $n(S \cap B \cap Z) = x$.
Then as $n(Z \cap S) = 10$ you can insert the $10 - x$; similarly for $6 - x$ and $8 - x$.
See Figure 3:14.

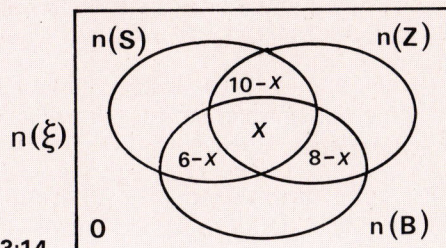

Fig. 3:14

The fourth section of each set can now be completed, e.g. $n(S) = 17$ and so far $n(S) = 10 - x + x + 6 - x = 16 - x$; hence writing $1 + x$ in the last section will give the total 17. See Figure 3:15.

Finally add up all the regions to total 32 and solve the resulting equation to find x. The diagram can then be easily completed: see Figure 3:16.

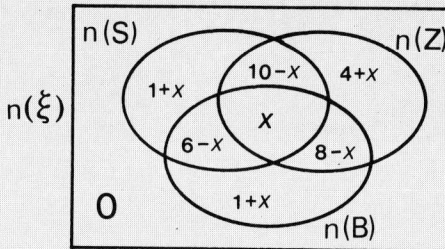

Fig. 3:15

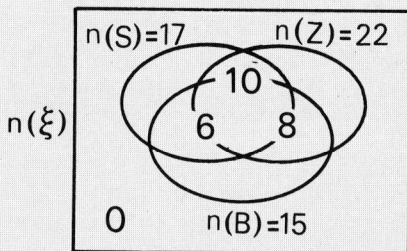

Fig. 3:16

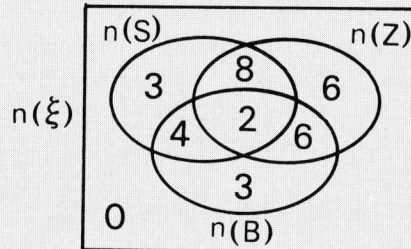

Fig. 3:17

Solution Method Two

See Figure 3:17. Compare the union of the sets with their sum: $n(S \cup B \cup Z) = 32$ and $n(S) + n(B) + n(Z) = 54$. The difference is 22, which means that we must 'lose' 22 in the intersections.

The given intersections total 24, which is 2 too many. This tells you that the intersection of all three contains 2. The diagram may now be easily completed: see Figure 3:16.

For Discussion

The diagram for part (a) may be filled in directly from the given information. Those for (b) and (c) need one of the above two methods.

(a) Of the households in a street, 2 took the Star, the News and the Post, 6 took (at least) the Star and the News, 4 took the Star and the Post, 5 took the News and the Post, 12 took the Star only, 20 took the News, 20 took the Post, 16 took none of these.

(b) Of 42 pupils, 14 had sisters, 25 had cousins, 17 had brothers, 8 had cousins and sisters, 7 had cousins and brothers, 6 had sisters and brothers, 1 belonged to none of these three categories.

(c) $\xi = \{\text{holidaymakers}\}$; $F = \{\text{those going to France}\}$; $S = \{\text{those going to Spain}\}$; $I = \{\text{those going to Italy}\}$; $n(\xi) = 97$; $n(F) = 35$; $n(S) = 47$; $n(I) = 42$; $n(F \cap S) = 20$; $n(F \cap I) = 13$; $n(S \cap I) = 21$; $n(F \cup S \cup I)' = 19$.

1   Figure 3:18 shows the result of a survey of the number of pupils who play hockey (H), cricket (C) and tennis (T). How many play:
(a) both hockey and tennis   (b) all three   (c) none   (d) only cricket
(e) hockey   (f) cricket   (g) hockey and/or cricket?

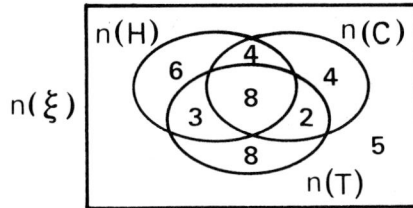

Fig. 3:18

*2   Figure 3:19 shows how class 4A travelled to school during a term. How many went:
(a) by bus   (b) by cycle   (c) on foot   (d) by bus, but not on foot
(e) by cycle but not by bus   (f) only on foot   (g) only by cycle
(h) by bus, but not by cycle   (i) by cycle or on foot or both, but not by bus
(j) by exactly two methods   (k) by at least two different methods
(l) by some other means than these three?

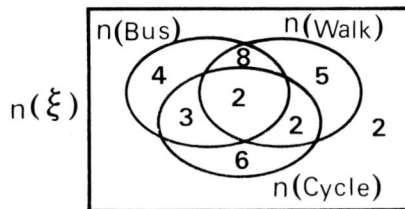

Fig. 3:19

*3   Figure 3:20 shows the activities that some friends took part in one week.
(a) What did Caryl do that Jane did not?
(b) Who did most activities?   (c) Who did none?
(d) What did Penny do that Gene did not?
(e) Who went dancing?   (f) Who went skating?
(g) Who did not go to judo?

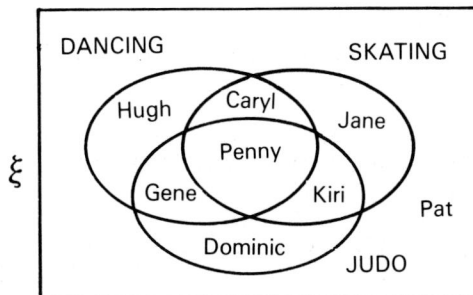

Fig. 3:20

*4 Draw a Venn diagram of three intersecting sets: J (judo), F (fencing) and S (squash). Complete it to show that in a club: 5 do judo, fencing and squash; 16 do judo and fencing only; 8 do judo and squash only; 3 do fencing and squash only; 26 do judo only; 7 do fencing only; 24 do squash only.

(a) How many do:
   (i) judo   (ii) fencing   (iii) squash?

(b) How many members in the club, if all do at least one of the three activities?

5 In a group of girls, 12 play hockey (H), 14 play netball (N), 10 play tennis (T), 4 play both netball and hockey but not tennis, 5 play netball only, 2 play hockey only, and 2 play both hockey and tennis but not netball.

Draw a Venn diagram to show this and state how many girls there are if all play at least one of the games.

6 All 30 boys in a class play at least one of soccer (S), rugby (R) and cricket (C). 18 play soccer, 17 play rugby, 6 play only soccer, 7 play only rugby, 3 play all three, and 8 play both soccer and cricket.

Draw a Venn diagram to show this and state how many play:
(a) both soccer and cricket, but not rugby   (b) soccer and rugby
(c) rugby but not cricket   (d) cricket.

7 Fifty boys turn out their pockets! 16 have a ruler, 8 a hanky, 36 a pen, 6 a ruler and a hanky, 3 a hanky and a pen, and 5 a ruler and a pen, and 3 have none of these. How many have:
(a) all three items   (b) only a pen?

8 A lady categorises men as tall (T), dark (D) and handsome (H). She observes 40 men and decides that 23 are tall, 26 are dark, 19 are handsome, 14 are tall and dark, 17 are tall and handsome, and 11 are dark and handsome, but 4 can only be described as short, grey and ugly!

Illustrate this (with a Venn diagram!) and say how many are:
(a) handsome but not dark   (b) tall, but not dark or handsome.

9 Illustrate with a Venn diagram sets A, B and C if:
$n(A \cup B \cup C)' = 0$, $n(A \cup B \cup C) = 28$, $n(A) = 10$, $n(B) = 16$, $n(C) = 20$,
$n(A \cap B) = 5$, $n(B \cap C) = 9$, $n(A \cap C) = 7$.

State:
(a) $n((B \cup C) \cap A)$ (b) $n(B \cup (C \cap A))$.

10 A, B and C are intersecting sets and:
$n(A \cup B \cup C)' = 0$, $n(B \cup C)' = 3$, $n(A \cup C)' = 1$, $n(B) = 6$, $n(A \cap B) = 5$,
$n(C) = 13$, $n(A \cap C) = 8$, $n(A \cup B \cup C) = 21$.

Find:
(a) $n(A \cap B \cap C)$ (b) $n(A)$.

11 Out of 42 cars, 23 had fog-lights, 30 had reversing-lights, 23 had spotlights, 8 had both fog- and spot- but not reversing-lights and 1 had only a spotlight. Thirty-two cars had at least two of these kinds of light and the number of cars having both fog- and reversing-lights but not spotlights was equal to the number having both spot- and reversing- but not fog-lights.

How many had:
(a) only reversing-lights   (b) none of these lights?

4 Calculators: powers, roots, reciprocals

● You need to know . . .

● How a calculator carries out calculations

All calculators have two kinds of keys:

Digits and the decimal point: 0 1 2 3 4 5 6 7 8 9 .

Function keys, e.g. $+$ $-$ $\times$ $\div$ x^2 $\sqrt{}$ $\frac{1}{x}$ $+/-$ TAN

Most calculators also have an $=$ key and memory, or store, keys, e.g. MS STO MR M+

It is important to remember that all function keys operate on the number showing in the display at the moment that they are pressed (but see the note on 'BODMAS' below). This is why $\frac{12}{2 \times 3}$ will come to the wrong answer if you key in 12 $\div$ 2 $\times$ 3 $=$. The correct answer is 2, because $12 \div 6 = 2$.

Key	12	$\div$	2	$\times$	3	$=$
Display	12	12	2	6	3	18

To get the right answer key in 12 $\div$ 2 $\div$ 3 $=$

Notice that the calculator does not appear to do anything when the first function key is pressed, but in fact it puts the display into a hidden memory called the y register. When the next function key is pressed the calculator carries out the first operation, combining the number in the y register with the number on display (x). It then displays the answer (new x) and also stores the answer in the y register. ('BODMAS' calculators do not always follow this system; this is explained later.) This is what happens when you work out $\frac{12}{2 \times 3}$:

Key	12	$\div$	2	$\div$	3		$=$
Display (x)	12	12	2	6	3		2
Hidden (y)	–	12	12	6	6		2

Most scientific calculators follow the BODMAS (Brackets; Of; Divide; Multiply; Add; Subtract) rule, in as much as they save up additions and subtractions until any multiplications and divisions have been carried out. Mathematicians consider that the correct answer to $4 + 3 \times 2$ is $4 + 6 = 10$, not $7 \times 2 = 14$.

Key in: 4 $+$ 3 $\times$

If your display is still 3, then your calculator is following the BODMAS rule; if it shows 7 it is not going to give the correct answer.

A BODMAS calculator stores up the pending operations in a 'stack', for which it has extra hidden memories. The example $4 + 3 \times 2$ only needs one extra memory (z):

Key	4	$\boxed{+}$	3	$\boxed{\times}$	2	$\boxed{=}$		
Display (x)	4	4	3	3	2	→ 6	→10	
Stack (y)	–	+4	+4	3	3			
Stack (z)	–	–	–	+4	+4	+4		

A similar effect is obtained by using bracket keys:
e.g. (4 + 3) × 2 = 14

Key	$\boxed{(}$	4	$\boxed{+}$	3	$\boxed{)}$	$\boxed{\times}$	2	$\boxed{=}$
Display (x)	–	4	4	3	3	3	2	→14
Stack (y)	–	–	+4	+3	7	7	7	
Stack (z)	–	–	–	+4	–	–	–	

● Constant functions (k)

Most calculators will remember the last operation and keep repeating it. Some do this always, others need two presses of the function key or the use of a k key. Try the following. One of them, at least, will probably work out 2 × 2 × 2 = 8.

2 $\boxed{\times}$ $\boxed{=}$ $\boxed{=}$ 2 $\boxed{\times}$ $\boxed{\times}$ $\boxed{=}$ $\boxed{=}$ 2 $\boxed{\times}$ $\boxed{k}$ $\boxed{=}$ $\boxed{=}$

Now try:

4 $\boxed{\div}$ 2 $\boxed{=}$ 6 $\boxed{=}$ 4 $\boxed{\div}$ 2 $\boxed{k}$ $\boxed{=}$ 6 $\boxed{=}$ 4 $\boxed{\div}$ $\boxed{\div}$ 2 $\boxed{=}$ 6 $\boxed{=}$

● Memory keys

There are two kinds of memory; one is called a 'store', the other is called an 'accumulator'.
Store keys are usually marked MS or STO or $x \to M$.
Accumulator keys are usually marked M+ or M+= or M– or M–= or ACC.

The memory content is recalled by a key usually marked RM or MR or REC.

Examples

Key	6	MS	8	MR	9	MS	MR
Display	6	6	8	6	9	9	9
Store	0	6	6	6	6	9	9

Key	6	M+	8	M+	9	MR	
Display	6	6	8	8	9	14	
Store	0	6	6	14	14	14	

You have to be careful to note whether or not the memory is empty when using the M+ key. Usually the display shows **M** to remind you that there is something in the memory. There is no problem with the MS store key, because the memory is automatically emptied as soon as this key is pressed. There are various ways of cancelling memories. Consult your handbook.

Test yourself

1   Calculate:
(a) 3.65 × 0.801   (b) 5.01 ÷ 6.27   (c) 0.0954 – 0.312

2 Calculate:

(a) $\dfrac{51\,580\,000}{4163 \times 9129}$ (b) $\dfrac{306.2}{1.681 \times 5}$ (c) $\dfrac{71.6 \times 305}{42.3 \times 8.6}$

(d) $\dfrac{1002 \times 363}{928 \times 15.6}$ (e) $\dfrac{3121 - 14.56}{431.6 \times 1.06}$ (f) $\dfrac{22.41 + 380.1}{5.1 \times 16}$

---

3   Calculate correct to 3 significant figures:

(a) $\dfrac{1.8 - 1.09}{7.6 \times 0.8154}$ (b) $\dfrac{18.15 \times 0.512}{6.631 \times 1.899}$ (c) $(14.26 \times 3.85) - (7.65 \div 91.3)$

(d) $(16 - 9.99) \div (1.284 \div 16.5)$ (e) $\sqrt{1.82} \times 16.5$ (f) $\sqrt{0.182} \times 16.5$

(g) $(0.652)^2 \times (1.763)^2$ (h) $\dfrac{(1.25)^2 + 1}{\sqrt{0.016}}$ (i) $\sqrt{\dfrac{4.215 \times 0.1668}{(3.59)^2 + (4.12)^2}}$

4   Invent some number games that use a calculator.

Powers; roots; reciprocals

Powers

Powers like 2^3 can be worked out using the constant facility explained on page 27, but scientific calculators usually have a y^x key[†], which is quicker if you have a large index, like 2^9.

Key	2	$\boxed{y^x}$	9	$\boxed{=}$
Display (x)	**2**	**2**	**9**	**512**
Store (y)	–	2	2	–

Roots

Use the $\sqrt{}$ key for square roots (and the $\sqrt[3]{}$ key for cube roots if you have one). For further roots you need either a $y^{\frac{1}{x}}$ key (sometimes marked $\sqrt[x]{y}$) or a y^x and $\dfrac{1}{x}$ key (but see question 4 in the exercise for a 'trial and error' method that only uses basic functions).

Remember that $8^{\frac{1}{2}}$ is an alternative way of writing $\sqrt{8}$, $8^{\frac{1}{3}} \equiv \sqrt[3]{8}$, and $8^{\frac{1}{4}} \equiv \sqrt[4]{8}$, so $y^{\frac{1}{x}} \equiv \sqrt[x]{y}$.

Example   To find $\sqrt[4]{2}$.

Key: 2 $\boxed{y^{\frac{1}{x}}}$ 4 $\boxed{=}$ or 2 $\boxed{y^x}$ 4 $\boxed{\frac{1}{x}}$ $\boxed{=}$ or 2 $\boxed{\text{INV}}$ $\boxed{y^x}$ 4 $\boxed{=}$

Reciprocals

The $\dfrac{1}{x}$ (or 1/x) key is the **reciprocal** key. The reciprocal of x is the fraction $\dfrac{1}{x}$, usually changed to a decimal. The reciprocal of 4 is 0.25.

[†]Labelled x^y on Casio calculators.

‍

Reciprocals of fractions sometimes cause trouble to students, e.g. the reciprocal of $\frac{3}{4}$ is $\frac{1}{\frac{3}{4}}$.

Think of this as $1 \div \frac{3}{4} \rightarrow 1 \times \frac{4}{3} = 1.\dot{3}$. Your calculator would need the key sequence

$3 \boxed{\div} 4 \boxed{=} \boxed{\frac{1}{x}}$.

Remember that a function key operates on the number in the display. Function keys like x^2, $\sqrt{}$, TAN, etc. operate as soon as they are pressed. Others, like $\times$, $+$, y^x, etc. wait until another function key or = is pressed before operating; this is because they combine the x and y registers.

To find the reciprocal of $\frac{3}{4}$ it would be wrong to key in $3 \boxed{\div} 4 \boxed{\frac{1}{x}} \boxed{=}$. Why?

Reciprocals are 'self-inverses'. Try $4 \boxed{\frac{1}{x}}\boxed{\frac{1}{x}}\boxed{\frac{1}{x}}\boxed{\frac{1}{x}}$.

In the following questions give all answers correct to 3 significant figures.

1 Calculate:
   (a) $(26.7)^3$ (b) $(13.2)^4$ (c) $(1.06)^2$.

2 Calculate:
   (a) $\sqrt[3]{20}$ (b) $\sqrt[4]{150}$ (c) $\sqrt[3]{182}$ (d) $\sqrt[3]{9.6}$ (e) $\sqrt[3]{17.6}$ (f) $\sqrt[4]{2000}$.

3 Calculate the reciprocal of:
   (a) 1.56   (b) 7.26   (c) $\frac{1}{3}$ (d) $1\frac{1}{2}$ (e) $1\frac{3}{4}$.

---

4 The following method is useful if you do not have a y^x key, and also helps your understanding of decimals.

   Example To find $\sqrt[3]{35.6}$.

      As $3 \times 3 \times 3 = 27$ and $4 \times 4 \times 4 = 64$, $3 < \sqrt[3]{35.6} < 4$
      Try 3.3 as the answer: $3.3 \times 3.3 \times 3.3 = 35.937$, so $\sqrt[3]{35.6} < 3.3$
      Try 3.2: $3.2 \times 3.2 \times 3.2 = 32.768$, so $\sqrt[3]{35.6} > 3.2$
      Try 3.25: $3.25^3 = 34.328\,125$, which is too small.
      Try 3.27, etc. etc., say to 3.289\,652\,4 which is sufficiently exact!

   Repeat question 2 using this method.

5 **Using a calculator for non-decimal conversion**
   Most calculators cannot cope simply with units like hours, minutes, seconds; degrees, minutes; etc. This is because these units are not based on the tens system. The main problem is that a calculator always gives a remainder as a decimal, so that 130 min divided by 60 gives 2.16666667 instead of 2 h 10 min. We can recover the 10 min by subtracting the 2, then multiplying the decimal part by the number we divided by (60). Try it.

The rule (or 'algorithm') to obtain a remainder to a division is: 'Subtract the integral part of the answer, then multiply by the divisor'. (There may be a small error, due to 'rounding errors', e.g. 2.999998 for remainder 3 is clearly the calculator's error.

Example   To convert 25 086 s to hours and minutes.

Key	25 086	÷	60	=	−	418	=	×	60	=
Display	**25 086**		**60**	**418.1**		**418**	**0.1**		**60**	**6**

Write down 418 min
(or save in memory)

This is the
6 s remainder.

(or recall memory)

Key	418	÷	60	=
Display	**418**		**60**	**6.9666667**

This is the hours.
Write down 6 h.

Key	−	6	=	×	60	=
Display		**0.9666667**			**60**	**58**

These are the remaining minutes.

Answer: 25 086 s = 6 h 58 min 6 s.

Work the above example, making sure you really understand the reason for each step, then change:

(a) to hours, minutes and seconds:
    (i) 31 012 s   (ii) 43 161 s   (iii) 100 301 s   (iv) 89 135 s
    (v) 30 016 s   (vi) 210 193 s.
(b) to degrees and minutes (1° = 60′) to the nearest minute:
    (i) 46.8°   (ii) 60.25°   (iii) 41.87°   (iv) 163.76°
(c) to decimals of a degree, correct to 3 s.f.:
    (i) 32° 16′   (ii) 21° 5′   (iii) 80° 43′   (iv) 18° 54′

Note: Many scientific calculators have a special key for the conversions in parts (b) and (c). This key can also be used for part (a).

(d) to yards, feet and inches (1 yd = 3 ft, 1 ft = 12 ins):
    (i) 6102 ins   (ii) 8134 ins   (iii) 10 000 ins.

5 Triangles: congruency; similarity

● You need to know . . .

● Straight-line and parallel-line angles

In Figure 5:1 angle *a* is **acute**
     angle *b* is **obtuse**
     angle *r* is **reflex**.

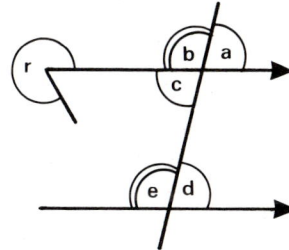

$a + b = 180°$ (**Adjacent** ∠**s on a straight line**)
  $a = c$ (**Vertically opposite** ∠**s**)
  $a = d$ (**Corresponding** ∠**s**)
  $c = d$ (**Alternate** ∠**s**)
$c + e = 180°$ (**Allied** ∠**s**) [Also known as Interior ∠s]

Fig. 5:1

Hints: Adjacent ⇒ next-door
   Corresponding ⇒ in the same position
   Alternate ⇒ on opposite sides (Z angles)
   Allied ⇒ joined together

Angles that add up to 180° are said to be **supplementary**.

● Construction methods

Figure 5:2 shows the compass construction for a 60° angle.

Figure 5:3 shows the way to bisect an angle.

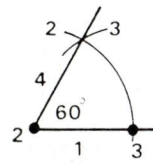

Fig. 5:2

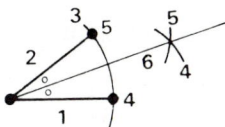

Fig. 5:3

These two constructions can be combined to draw many other angles, e.g. 30° by bisecting 60°.

Figures 5:4(a) and 5:4(b) show two methods to construct a right angle. The first combines a 60° and a 30° angle; the second bisects a 180° angle.

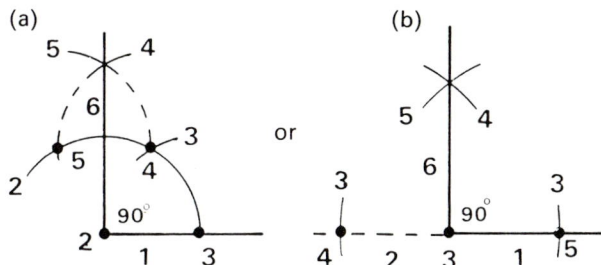

Fig. 5:4

Figures 5:5 to 5:11 show other constructions often required at a 16+ examination.

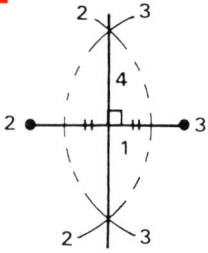

Fig. 5:5 Perpendicular bisector

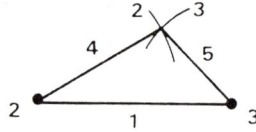

Fig. 5:6 △ given 3 sides

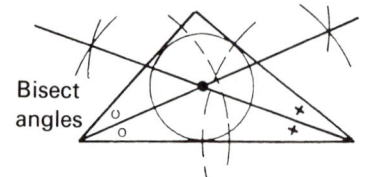

Bisect angles

Fig. 5:7 Incircle of △

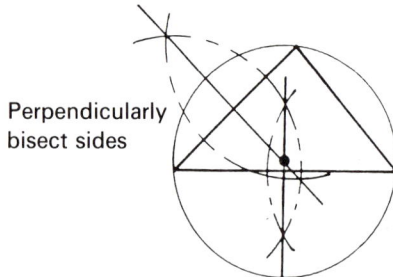

Perpendicularly bisect sides

Fig. 5:8 Circumcircle of △

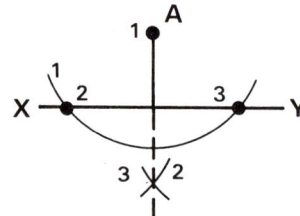

Fig. 5:9 Dropping a perpendicular

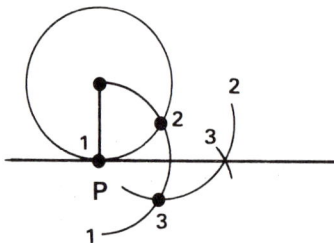

Fig. 5:10 Tangent to a circle

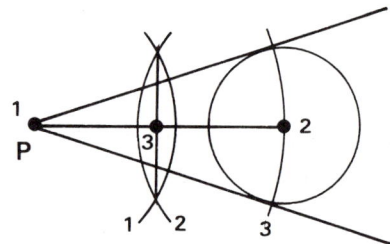

Fig. 5:11 Two tangents from a point

● **Names and properties of special quadrilaterals**

Name	Sides	Angles	Diagonals
Trapezium	1 pr //	—	—
Isosceles trapezium	1 pr //, 1 pr non-// equal	2 equal prs	equal
Kite	2 adjacent equal prs	1 equal pr	one bisected, cross at 90°
Parallelogram	2 prs equal and //	opposites equal	bisect each other
Rhombus	4 equal	opposites equal	bisect at 90°
Rectangle	as parallelogram	all 90°	equal
Square	as rhombus	all 90°	equal, cross at 90°

// = parallel

Test yourself

1 Work with a partner, testing each other on the names and facts about straight-line and parallel-line angles.

2 Practise the constructions. Check with your teacher which constructions have to be learnt for your examination.

3 Draw accurate diagrams of each of the special quadrilaterals and check the properties by measurement and observation.

A Congruent triangles

Congruent figures are exactly the same shape and size.

There are four 'cases of congruency' for triangles. In each case, three equalities must be known.

One   3 sides

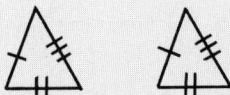

Two   2 sides, included angle

(*Included* means 'between the sides'.)

Three   Right angle, hypotenuse, side

(The *hypotenuse* is the side opposite the right angle.)

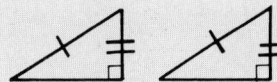

Four   2 angles, corresponding side

(*Corresponding* means in the same position: opposite the same angle.)

Example   In Figure 5:16, Δs $\frac{ABC}{CDA}$ are congruent (3 sides).

Note: AC is the third side. It is **common** to both triangles, therefore equal in both. We mark a common side with a wavy line.

From $\frac{ABC}{CDA}$: covering up $\frac{A}{C}$ gives BC = DA

covering up $\frac{B}{D}$ gives AC = CA

covering up $\frac{C}{A}$ gives AB = CD.

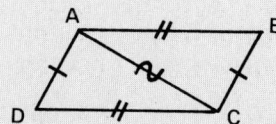

Fig. 5:16

5

1 Where the following pairs of triangles are congruent, write the equal angles and sides over each other and give the case of congruency. If the triangles are not congruent, explain briefly why not.

Note: The triangles are often drawn deliberately to appear not congruent when they are, and vice versa. Only take sides and angles as equal if they are marked equal.

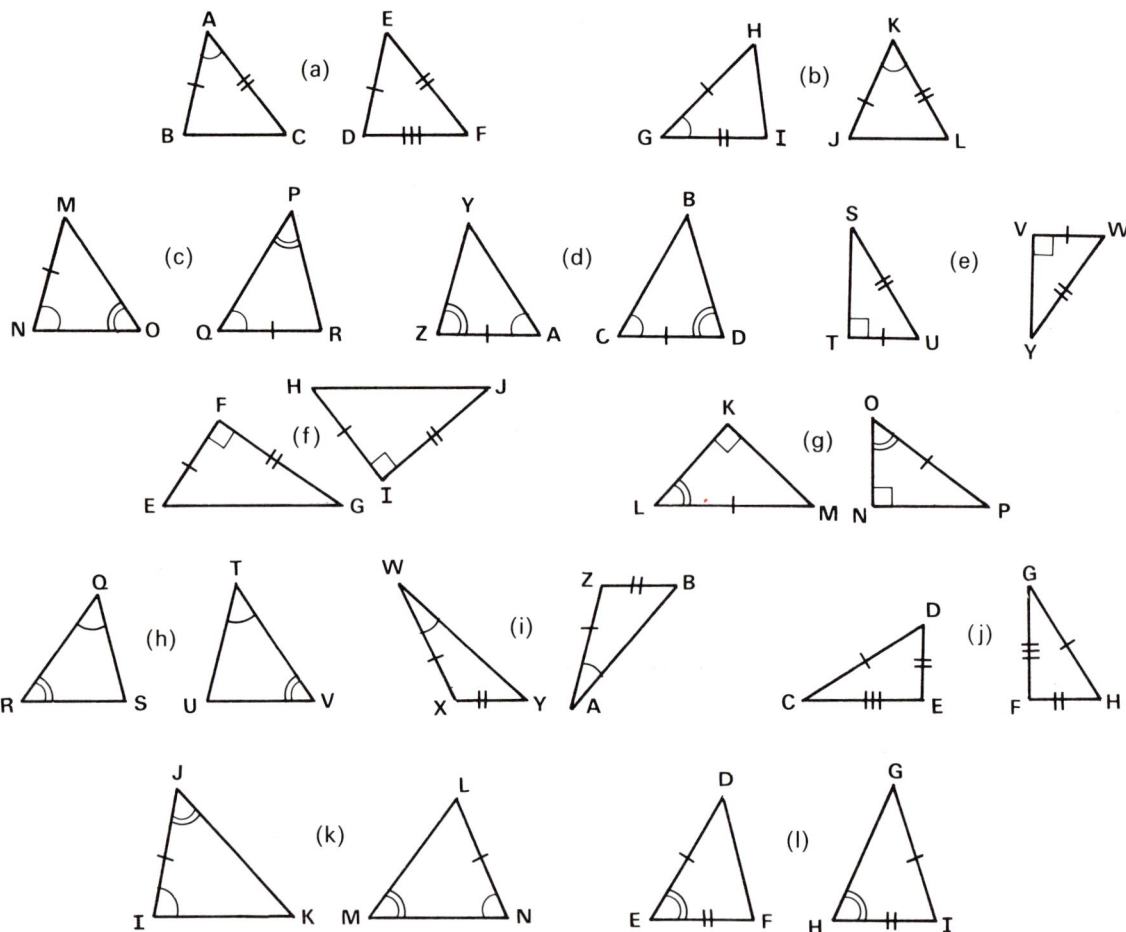

Fig. 5:17

*2 Construct accurately the following pairs of triangles. It will help if you roughly sketch the triangles first. Say why each pair are or are not congruent.

(a) AB = 3 cm; BC = 4 cm; AC = 2 cm
 DE = 2 cm; EF = 3 cm; DF = 4 cm

(b) HI = 4 cm; ∠H = 30°; HG = 5 cm
 JK = 5 cm; ∠K = 30°; KL = 4 cm

(c) ON = 6 cm; ∠O = 60°; NM = 5.5 cm
 QR = 6 cm; ∠Q = 60°; PQ = 5.5 cm

(d) SU = 3 cm; ∠U = 90°; ST = 5 cm
 VW = 3 cm; ∠V = 90°; WX = 5 cm

(e) AZ = 4 cm; ∠Z = 60°; ∠A = 45°
 BC = 4 cm; ∠C = 60°; ∠D = 45°

(f) QR = 5 cm; ∠Q = 45°; SR = 4 cm
 TU = 5 cm; ∠T = 45°; US = 4 cm

(g) IJ = 4 cm; ∠J = 60°; HI = 3 cm
 AG = 4 cm; ∠G = 60°; AE = 3 cm

3 Give reasons for the following pairs of triangles being congruent or not. Where they are congruent, write the equal angles over each other, as in question 1.

(a)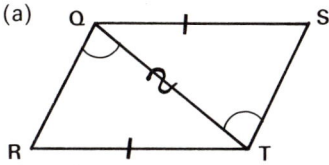

(b)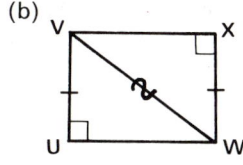

(c)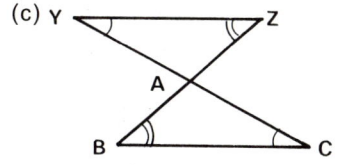

(d)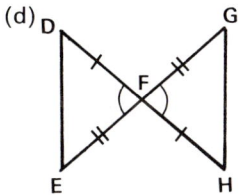

(e)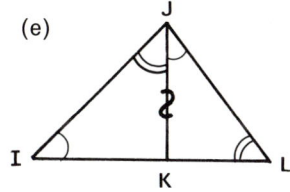

(f)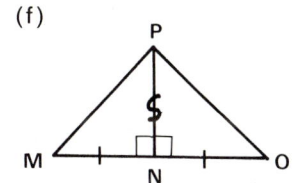

Fig. 5:18

4 **Example** *Given* Figure 5:19, in which AB // DC and AB = DC.
To prove BC // AD, making ABCD a parallelogram.

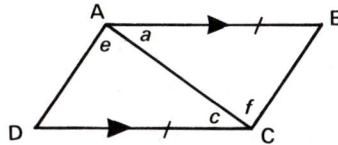

Fig. 5:19

Proof In △s ABC and ADC,
　　AB = DC (given)
　　AC is common
　　$a = c$ (alternate ∠s, AB // DC)
∴ △s $\frac{ABC}{CDA}$ are congruent (2 sides, included angle)
∴ $e = f$ (∠s of congruent triangles)
∴ AD // BC (*e* and *f* are alternate *and* equal)

(a) Prove that in Figure 5:20 EF = HG and EH = FG.

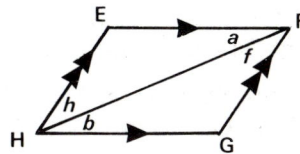

Fig. 5:20

(b) Prove that in Figure 5:21 IJ // LM.

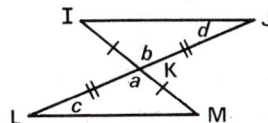

Fig. 5:21

(c) In Figure 5:22 ABCD and BEGF are
squares.
  (i) Why must ∠ABE = ∠FBC?
  (ii) Prove that △s ABE and CBF are
     congruent.
  Note: This is part of the Euclidean
  proof of Pythagoras' Theorem.

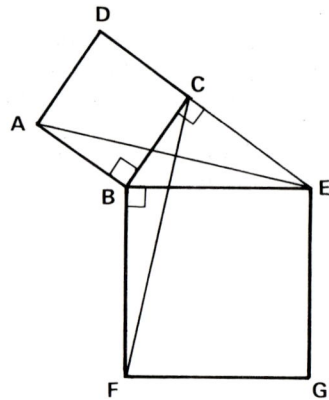

Fig. 5:22

5 In △ABC, AB = AC. AD is drawn perpendicular to BC, D lying on BC.

(a) Prove that ∠B = ∠C.

(b) State what other equalities are true in ∠ABC.

6 Assuming nothing more about a parallelogram than that it is a quadrilateral with both pairs of opposite sides parallel, prove that the opposite sides and opposite angles of a parallelogram are equal.

7 Prove that a quadrilateral with both pairs of opposite sides equal must be a parallelogram.

8 Find the rest of the proof of Pythagoras' Theorem referred to in question 4.

B Similar triangles

Two shapes are similar when one is an exact enlargement of the other.

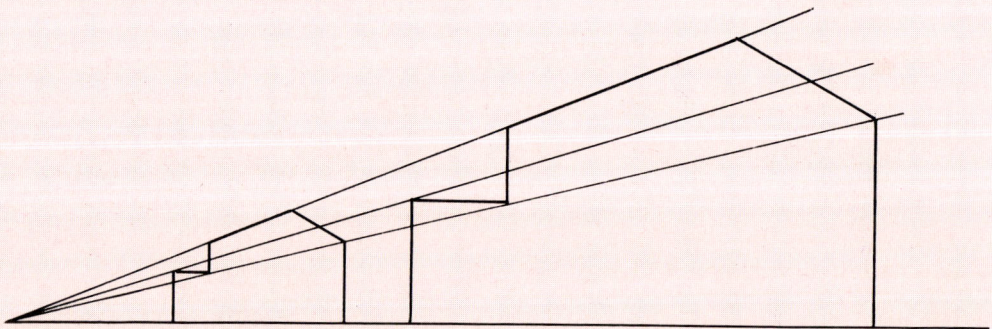

Fig. 5:23

When two shapes are similar, their corresponding (in the same position) sides are in the same ratio, and their angles are the same sizes. For a triangle, and only for a triangle, it is sufficient to know *one* of these two facts to know that the shapes are similar.

In Figure 5:24 the two shapes have angles of the same sizes, but they are not similar.

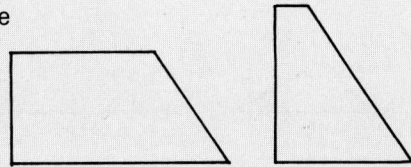

Fig. 5:24

In Figure 5:25 the two shapes have their sides in the same ratio (2 : 1) but they are not similar.

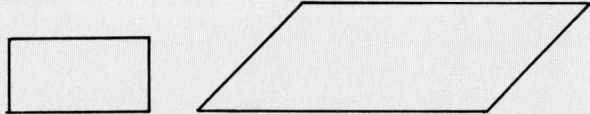

Fig. 5:25

The triangles in Figure 5:26 are similar because their angles are the same sizes. By writing the equal angles over each other, $\begin{smallmatrix} A & B & C \\ D & E & F \end{smallmatrix}$, then covering up one column at a time we obtain the ratio of the sides: $\dfrac{BC}{EF} = \dfrac{AC}{DF} = \dfrac{AB}{DE}$.

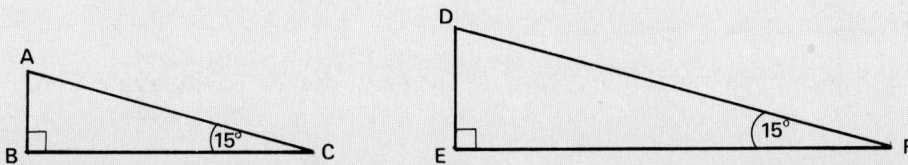

Fig. 5:26

Example Calculate the unknown lengths in Figure 5:27.

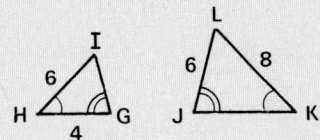

Fig. 5:27

Δs $\begin{smallmatrix} G & H & I \\ J & K & L \end{smallmatrix}$ are similar, $\therefore \dfrac{HI}{KL} = \dfrac{GI}{JL} = \dfrac{GH}{JK}$.

Substituting the given lengths we have: $\dfrac{6}{8} = \dfrac{GI}{6} = \dfrac{4}{JK}$.

The sides are therefore in the ratio 6 : 8 = 3 : 4.
The sides of the smaller triangle are $\frac{3}{4}$ of the sides of the larger.
The sides of the larger triangle are $\frac{4}{3}$ of the sides of the smaller. Hence

$$GI = JL \times \frac{3}{4} = \overset{3}{\cancel{6}} \times \frac{3}{\cancel{4}_2} = 4\tfrac{1}{2} \text{ cm}$$

$$JK = GH \times \frac{4}{3} = 4 \times \frac{4}{3} = 5\tfrac{1}{3} \text{ cm}.$$

1 For each pair of similar triangles in Figure 5:28
   (i) write the equal angles over each other
   (ii) obtain the equation of the sides
   (iii) substitute for the sides given and find the two unknown sides. All dimensions are in cm.

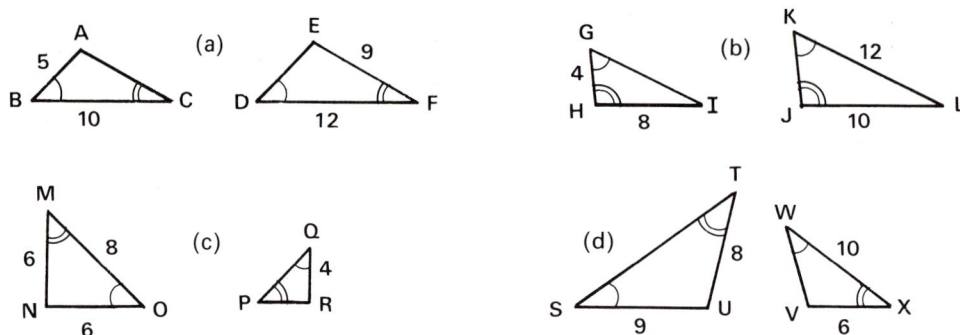

Fig. 5:28

*2 Calculate the unknown sides in the pairs of similar triangles in Figure 5:29.

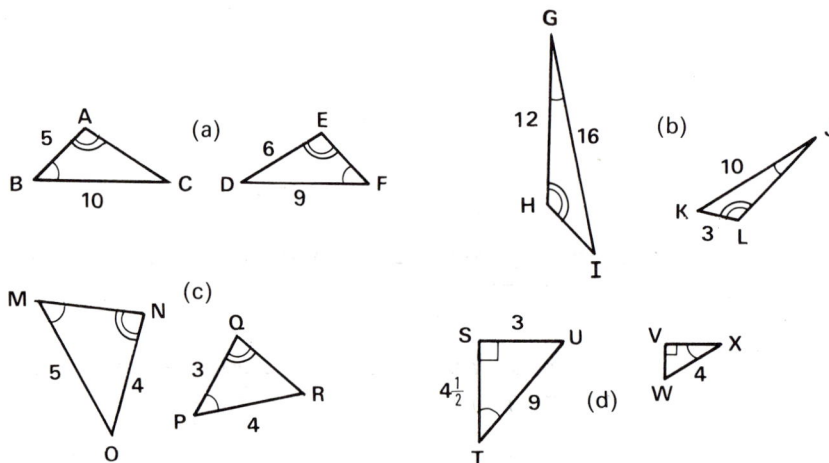

Fig. 5:29

3 Calculate by similar triangles the unknown sides in Figure 5:30.

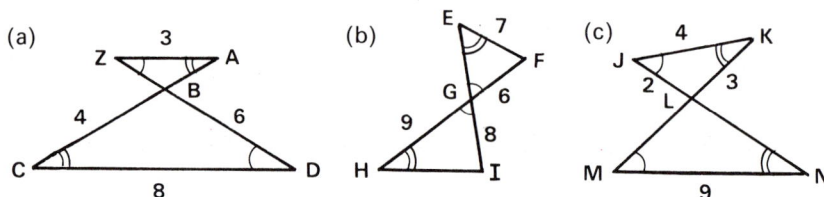

Fig. 5:30

4 In Figure 5:30(a), why must ZA be parallel to CD?

5 Copy Figure 5:31. Mark the angles that you know must be equal, then calculate the unknown lengths.

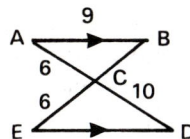

Fig. 5:31

6 Copy Figure 5:32. Mark the angles that you know must be equal, then calculate all unknown lengths.

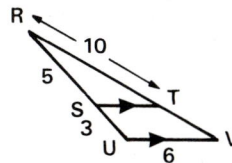

Fig. 5:32

7 Figures 5:33 and 5:34 illustrate the Midpoint Theorems.

In Figure 5:33, '**The line from the midpoint of one side of a triangle, parallel to another side, bisects the third side.**'

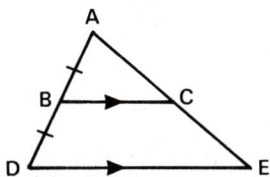

Fig. 5:33

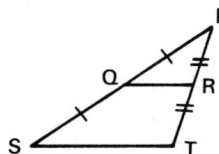

Fig. 5:34

In Figure 5:34, '**The line joining the midpoints of two sides of a triangle is parallel to, and a half of, the third side.**'

Draw, any size, copies of Figures 5:33 and 5:34, then check by measurement that the two theorems are true for *your* diagrams. (The formal proofs are a little involved, but you will find them in a standard geometry text-book.)

8 In a triangle ABC, AB = 6 cm, BC = 8 cm, and AC = 7 cm. X is the midpoint of AB, and Y is on AC such that XY // BC.

(a) Calculate the lengths of AY and XY.

(b) Show that $\triangle$s AXY and ABC are similar, and that $\dfrac{AX}{AB} = \dfrac{AY}{AC} = \dfrac{XY}{BC}$.

9 WXYZ is a quadrilateral. P, Q, R and S are the midpoints of the sides. Show that PQRS is a parallelogram.

10 ABCD is a kite. P, Q, R and S are the midpoints of the sides. Show that PQRS is a rectangle.

Using your calculator

Piston speeds

(*Shell Photo Service*)

'You're cruising on the motorway at 70 m.p.h. At the same time, parts of your engine are covering greater distances than the car. Each minute your pistons race up and down nearly 3500 times. The four pistons travel a total of about $1\frac{3}{4}$ times the distance covered by the average car. Yet only millionth fractions of a centimetre separate the pistons from the cylinder.'

Use the above information and your calculator to answer the following questions.

1 What product do you think this could be advertising?

2 How many times do the pistons move up and down in a second?

3 In the same gear as described in the advertisement, how many times will the pistons move up and down per minute at:
(a) 35 m.p.h.   (b) 60 m.p.h.   (c) 100 m.p.h.?

4 If the car travels 60 miles, how many miles will the pistons have travelled altogether, up and down?

5 At 70 m.p.h., how many times do the pistons move up and down per mile?

6 Using your answer to question 5, and the fact that the four pistons' total travel amounts to about $1\frac{3}{4}$ times the distance covered, what is the length of stroke of a piston in cm? (That is, the distance it moves one way up the cylinder. Take 5 miles = 8 km.)

6 Decimal fractions: standard form

● You need to know . . .

● Approximation methods

Many amounts used in life are approximations to (not exactly) the true amount. This may be because there is no exact amount (e.g. the length of a line), or to make the number easier to read or remember (e.g. a football crowd of 21 000).

Approximations may be expressed in many ways, e.g. 15.79 is 16 **to the nearest whole number**; 31 215 is 31 200 **to the nearest hundred**; 7.68 cm is 7.7 cm **to the nearest mm**; £10 ÷ 3 is £3.33 **to the nearest penny**.

You will often need to approximate calculator answers. How approximate you should make them usually depends on the information supplied, e.g. if a question is based on an average speed correct to the nearest km/h, and a distance correct to the nearest km, then it is silly to give a time for the journey correct to the nearest second, and even sillier to give an answer like 1.245367 hours.

Two special approximations are used in mathematics:

Decimal places (d.p.)
This states the number to a given number of figures after the decimal point. Clearly it is of no use when there are no figures after the point!

Examples 7.0145 → 7.015 to 3 d.p.

7.0145 → 7.01 to 2 d.p.

7.98   → 8.0 to 1 d.p. (the 'key' 8 makes 9 → 10)

Significant figures (s.f.)
All figures are counted, except zeros between the decimal point and the first non-zero digit, and place-value zeros before the point.

Examples 126.87 → 130 to 2 s.f. (The zero is not a significant figure, but it is needed to show the empty units' column, otherwise 126.87 → 13, which is silly.)

0.001 34 → 0.0013 to 2 s.f.

0.0598   → 0.060 to 2 s.f.

It is very important to check that your approximated answer *is* approximately the same size as the original number.

Test yourself

1 (a) By rounding each number to the nearest ten, find 'in your head' the approximate sum of:
16, 18, 23, 38, 40, 27, 8, 14 and 29.

(b) Find the accurate answer to (a).

2 As I go round the supermarket with my shopping list I roughly add up the costs to the nearest 50p so that I can check that I have enough cash to pay the bill. Compare my approximation with the correct bill if the goods cost:
35p, 82p, £1.14, £2.31, 15p, 15p, 61p, 73p, 67p, 82p, 50p, 97p, 75p.

3 Make up some questions like 1 and 2 above, comparing the approximate answers with the true ones. Or, even better, obtain some 'real' data, e.g. marks on a series of tests and some supermarket check-out till slips.

4 Write the following numbers correct to 2 d.p.
(a) 17.138   (b) 0.777   (c) 1.01282   (d) 10.1059   (e) 19.1051   (f) 206.796

5 Write the numbers in question 4 correct to 2 s.f.

6 If you need more practice it is easy to work in pairs, making up questions and testing each other.

7 What is rather strange about 9.99 when corrected to 2 s.f.?

---

8 Approximating with a computer
The INT function is used in BASIC when approximating. INT takes the integral part of a number and ignores the decimal part, so that $INT(2.48) = 2$ and $INT(0.81) = 0$.

If we want 3.4125 correct to 2 d.p. we first multiply by 100 to move the 2 decimal place figures in front of the point $(3.4125 \times 100 = 341.25)$. Then we use the INT function to 'lose' the rest of the number $(INT(341.25) = 341)$. Finally divide by 100 to restore the figures' true place value.

Here it is in BASIC:

```
10 LET A = 3.4125
20 LET B = 3.4125 * 100
30 LET C = INT(B)
40 LET D = C/100
50 PRINT A; "ˏcorrect to 2 d.p. isˏ"; D
```

By combining lines 20 to 50 we can shorten the program to:

```
10 LET A = 3.4125
20 PRINT A; "ˏcorrect to 2 d.p. isˏ"; (INT(3.4125 * 100))/100
```

The above program would not be correct if A = 3.4156, as it would fail to take account of the 'key' figure 5. We allow for this by increasing the key figure, whatever it is, by 5. This will automatically round-up the preceding figure if the key figure is 5 or more:

```
    3.4156    and      3.4125
+ 0.005              + 0.005
  3.4206               3.4175
```

There is a further problem when a 9 is involved, as the computer would drop any zeros after the point so that 4.7961 would be printed as 4.8 to 2 d.p., not 4.80, and 4.998 would be printed as 5, not 5.00.

In the following program lines 50 and 60 correct this fault. Work out how they do it.

```
 5 REM "2DP"
10 PRINT "Please type your number"
20 INPUT A
30 LET B = (INT((A + 0.005) * 100))/100
40 PRINT A; " correct to 2 d.p. is "; B;
50 IF B − INT(B) = 0 THEN PRINT ".0";
60 IF B * 10 − INT(B * 10) = 0 THEN PRINT "0"
```

When you fully understand the above program, try this one:

```
 5 REM "ANYDP"
10 PRINT "Please type your number"
20 INPUT A
30 PRINT "To how many places of decimals?"
40 INPUT B
50 IF LEN(STR$(A)) − LEN(STR$(INT(A))) < B + 2 THEN GOTO 10
60 LET C = INT(10 ↑ B) (Use ** on some computers)
70 LET D = (INT((A + 5/(C * 10)) * C))/C
80 PRINT A; " to "; B; " d.p. is "; D;
90 IF D − INT(D) = 0 THEN PRINT ".0";
100 IF B = 1 THEN GOTO 150
110 FOR F = 2 to B
120 LET D = D * 10
130 IF D − INT(D) < 0.0001 THEN PRINT "0";   (0.0001 instead of 0 allows for
140 NEXT F                                      inaccurate computer arithmetic)
150 PRINT
160 GOTO 10
```

Note
BBC computers
switch to standard
form when A < 0.1

9 Significant figure approximation can be done in a similar way, but as you have to allow for the original length of the number, for leading zeros, and for necessary place value zeros before the point, it is *much* more difficult. Do you like a challenge?

A   Standard form: numbers above 1

$$1.2346 \quad 08$$

This shows the way most scientific calculators display the answer to 123 456 × 1000. Because the answer (123 456 000) is too long for the display, it has been switched to **standard form**. The calculator we used cuts off all figures after the first five, so the number has also been rounded to 5 s.f. Yours may show more, or less, figures.

The 08 at the right is called the **exponent**. It tells you that 1.2346 is 8 columns too small. Moving the figures up the 8 columns gives the answer as 123 460 000, which is the most accurate this calculator can achieve. You may find it easier to think of the 08 as meaning that there are 8 figures between the first figure and the decimal point.

6

When we handwrite standard form we use the form $A \times 10^n$, where A is between 1 and 10 and n is an integer.

$$123\,456\,000 \to 1.234\,56 \times 10^8 \quad \text{(i.e. } 1.234\,56 \times 100\,000\,000\text{)}$$
$$318 \to 3.18 \times 10^2 \quad \text{(i.e. } 3.18 \times 100\text{)}$$

A calculator would not normally use standard form for numbers like 318, but using the $\boxed{\text{EXP}}$ or $\boxed{\text{EE}}$ key you can type in 318 as 3.18 $\boxed{\text{EXP}}$ 2 to give the display 3.18 02. When you type $\boxed{=}$ the calculator will probably switch it back to 318. If it does not, try typing $\boxed{\times}$ 1 $\boxed{=}$ instead.

Computers also use standard form for very large numbers (how large depends on your computer), but they show the exponent by an E, without leaving a gap, so 1.234 06 becomes 1.234E6.

1 Each of the following shows a number in standard form. Copy them, replacing the * by the correct number.
(a) $426 \to * \times 10^2$ (b) $* \to 2.75 \times 10^2$ (c) $3128 \to 3.128 \times *$ (d) $4000 \to 4 \times *$

2 Write in standard form:
(a) 200   (b) 5000   (c) 38   (d) 476   (e) 5900.

3 Write as ordinary numbers each of the following. Note that $10^1 \equiv 10$ ($\equiv$ means 'is identical to').
(a) 1×10^3 (b) 7.6×10^1 (c) $4.02\ \ 01$ (d) $7.6\ \ 02$ (e) 3.65×10^4
(f) 8.012E2   (g) $9.6\ \ 05$ (h) 3.14×10^3

4 Often standard form is combined with approximation.

Examples $319 \to 3.19 \times 10^2 \to 3.2 \times 10^2$ correct to 2 s.f.
$4000 \to 4.0 \times 10^3$ correct to 2 s.f.

Copy the following table, replacing all given distances by numbers expressed in standard form correct to 2 s.f.

	Mean distance from Sun in km	Diameter in km
Sun	–	1 392 000
Mercury	57 900 000	4 880
Venus	108 200 000	12 104
Earth	149 600 000	12 756
Mars	227 900 000	6 787
Jupiter	778 300 000	142 800
Saturn	1 427 000 000	120 000
Uranus	2 869 600 000	51 800
Neptune	4 496 600 000	49 500
Pluto	5 900 000 000	6 000

5   In standard form, $400 \rightarrow 4 \times 10^2$, and $40 \rightarrow 4 \times 10^1$. Write 4 in standard form.

6   **Example** $3.4 \times 10^2 \times 8 \times 10^5 = 27.2 \times 10^7 \rightarrow 2.72 \times 10^8$ in standard form.

Write in standard form:
(a) $4.9 \times 10^3 \times 5 \times 10$ (b) $7.6 \times 10^2 \times 3 \times 10^3$
(c) $8.9 \times 10^4 \times 2 \times 10^2$ (d) $3.1 \times 10^5 \times 2.1 \times 10^5$
(e) $4.12 \times 10^3 \times 8 \times 10^4$ (f) $6.3 \times 10^2 \times 7 \times 10^3 \times 2 \times 10^8$.

7   What is the smallest and the largest number that could be written as:
(a) 1.0×10^6 (b) 1.00×10^6 (c) 1.000×10^6 (d) 1.0000×10^6
(e) $1.000\,00 \times 10^6$?

8   Use a scientific calculator to work question 6.

B   Standard form: numbers below 1

7.8   -02

The calculator display here shows 0.078 in standard form. The -02 tells us that the 7.8 should be shifted two columns to the right to give the true value.

This would be handwritten as 7.8×10^{-2}.

10^{-2} is the index way of writing the fraction $\dfrac{1}{10^2} = \dfrac{1}{100}$.

1   Each of the following shows a number in standard form. Copy them, replacing the * with the correct number.
(a) $0.082 \rightarrow * \times 10^{-2}$ (b) $0.0046 \rightarrow * \times 10^{-3}$ (e) $0.065 \rightarrow 6.5 \times *$ (d) $0.46 \rightarrow 4.6 \times *$

2   Write as ordinary numbers:
(a) 8.06×10^{-2} (b) $6.09 \quad -01$ (c) 8×10^{-3} (d) 4.53×10^{-4} (e) $1.35 \quad -04$

3   Lyons Original Ready Brek has the following typical nutritional composition per 100 g.

Energy 1678 kJ
Protein 12.2 g
Fat 8.6 g
Carbohydrate 72 g
Vitamins: A, 0.0012 g   B1, 0.0018 g   B2, 0.0023 g
          C, 0.042 g   D, 0.000 017 g
Niacin 0.027 g   Iron 0.02 g   Calcium 1.2 g

Copy these details, but write each number in standard form.

*4 Change the following to standard form, to the stated degree of accuracy.
(a) 786; 1 d.p.   (b) 8000; 2 s.f.   (c) 399; 1 s.f.   (d) 24.989; 2 d.p.
(e) 103.97; 3 s.f.   (f) 60 991; 2 s.f.   (g) 31.7×10^2; 3 s.f.   (h) 431.6×10; 1 d.p.
(i) 39.97×10^5; 2 d.p.

*5 Write in standard form:
(a) 0.764 correct to 2 s.f.   (b) 0.009 85 correct to 2 s.f.

6 **Example** $\dfrac{6 \times 10^2 \times 2 \times 10^{-5}}{5 \times 10^3} = \dfrac{12 \times 10^{-3}}{5 \times 10^3} = 2.4 \times 10^{-6}$

Notes: (i) Add indices when multiplying: $10^2 \times 10^{-5} = 10^{-3}$.
(ii) Subtract indices when dividing: $10^{-3} \div 10^3 = 10^{-6}$.

Find in standard form:
(a) $4.1 \times 10^{-3} \times 2 \times 10^2$ (b) $8.2 \times 10^{-2} \times 2 \times 10^2$
(c) $1.2 \times 10^{-4} \times 1.2 \times 10^5$ (d) $4.1 \times 10^{-3} \times 3 \times 10^{-1}$
(e) $\dfrac{4 \times 10^2}{2 \times 10^{-3}}$ (f) $\dfrac{2 \times 10^{-5}}{5 \times 10^{-6}}$ (g) $\dfrac{7 \times 10^4}{28 \times 10^{-2}}$
(h) $\dfrac{9 \times 10^4}{4.5 \times 10^5}$ (i) $\dfrac{6.2 \times 10^{-3} \times 2 \times 10^2}{4 \times 10^{-2}}$
(j) $\dfrac{3.2 \times 10^{-8} \times 5 \times 10^{-6}}{4.8 \times 10^{-9}}$

---

7   As $10^{-3} = 1 \times 10^{-3} = 0.001 = \dfrac{1}{1000} = \dfrac{1}{10^3}$,

it follows that $10^{-2} = \dfrac{1}{10^2}$; $10^{-1} = \dfrac{1}{10}$; etc.

Similarly $4^{-1} = \dfrac{1}{4}$ and $4^{-2} = \dfrac{1}{4^2} = \dfrac{1}{16}$.

Write as fractions:
(a) 3^{-1} (b) 3^{-2} (c) 16^{-1} (d) 2^{-3} (e) 3^{-4} (f) 20^{-2}.

8   Explain clearly why 0.6×10^{-2} is not the same as 0.60×10^{-2}.

9   Explain clearly why $2 \times 10^2 \times 3 \times 10^3 = 6 \times 10^5$, but $2 \times 10^2 + 3 \times 10^3 \neq 5 \times 10^5$.

10   Use a standard form calculator to answer question 6.

● You need to know . . .

● How to transpose formulae ('change of subject')

$C = \pi d$ is the formula for the circumference of a circle; C is the **subject**.

When the formula is **transposed**, the subject is changed. The circumference formula can be transposed to make d the subject, giving $d = \dfrac{C}{\pi}$.

If the new subject-letter appears only once in the formula, the flow-diagram method illustrated below may be used. If the new subject-letter appears more than once, you have to use the 'balance' or 'change sides' algebraic method.

Examples (a) $u = s - t$; new subject s

$s \rightarrow -t \rightarrow u$
$s \leftarrow +t \leftarrow u$ giving $s = u + t$

(b) $u = s - t$; new subject t

$t \rightarrow -$ from $s \rightarrow u$
$t \leftarrow -$ from $s \leftarrow u$ giving $t = s - u$
Remember that '$-$ from' does not change.

(c) $p = sr$; new subject s

$s \rightarrow \times r \rightarrow p$

$s \leftarrow \div r \leftarrow p$ giving $s = \dfrac{p}{r}$

(d) $u = sr - t$; new subject s

$s \rightarrow \times r \rightarrow -t \rightarrow u$

$s \leftarrow \div r \leftarrow +t \leftarrow u$ giving $s = \dfrac{u + t}{r}$

Test yourself

1 Make t the subject of each of the following formulae.
(a) $a = t - g$ (b) $s = h - t$ (c) $f = 2t - y$

(d) $r = 4w - 3t$ (e) $w = ag + et$ (f) $k = \dfrac{a + 3t}{4}$

---

2 Make c the subject of each of the following formulae.
(a) $f = g - \frac{1}{2}(c + a)$ (b) $h = a^2 + c^2$ (c) $t = 2\pi\sqrt{c}$

(d) $d = ab(R - 2c)$ (e) $s = cr(a - t)$ (f) $f = \dfrac{3d - t}{c + b}$

---

3 Make r the subject of each of the following equations.

(a) $2g(r - p) + r = 3a$ (b) $r(2 + bc) = ar - st$

(c) $\dfrac{1}{r} - \dfrac{1}{t} = \dfrac{1}{a}$ (d) $\dfrac{2}{s} - \dfrac{b}{r} = 3a$ (e) $s = 3w\sqrt{\dfrac{2}{r}}$

A   Arcs (radians)

To find the length of an arc we first find what fraction of the circumference it is by considering the angle it subtends at the centre.

In Figure 7:1,

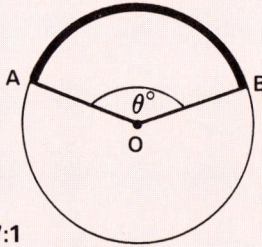

Fig. 7:1

$$\text{arc AB} = \frac{\theta^\circ}{360^\circ} \times \text{circumference} = \frac{\theta}{360} \times \pi \times \text{diameter}$$

Note: θ is a Greek letter, pronounced 'theta'.

Example   In Figure 7:2, $\theta = 50$ and the diameter is 36 cm.

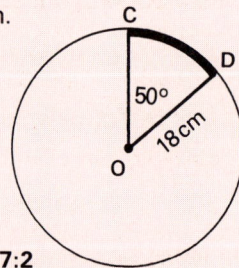

Fig. 7:2

$$\text{Arc CD} = \frac{50}{360} \times \pi \times 36 \to \frac{\overset{5}{\cancel{50}}}{\underset{1}{\cancel{360}}} \times \frac{22}{7} \times \cancel{36}^{1} = \frac{110}{7} \to 15\tfrac{5}{7}\text{ cm}$$

1   Find the length of each marked arc in Figure 7:3.

(a)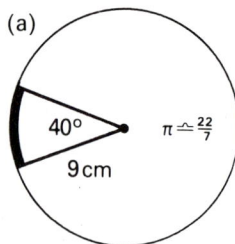

40° $\pi \simeq \frac{22}{7}$

9 cm

(b)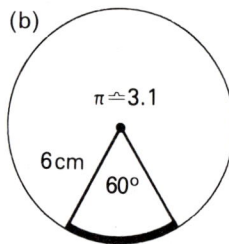

$\pi \simeq 3.1$

6 cm 60°

(c)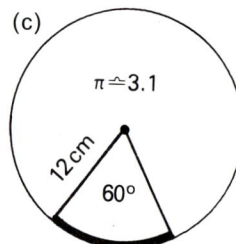

$\pi \simeq 3.1$

12 cm 60°

(d)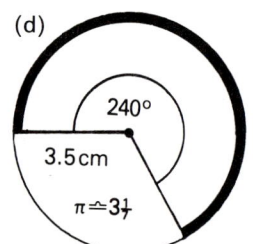

240°

3.5 cm

$\pi \simeq 3\tfrac{1}{7}$

Fig. 7:3

*2 State the values of (a) to (l) in this table.

Length of arc	60 cm	3 cm	20 cm	7 cm	10 cm	5 cm
Circumference	120 cm	12 cm	80 cm	21 cm	60 cm	25 cm
Arc as fraction of circumference	(a)	(c)	(e)	(g)	(i)	(k)
Angle at centre	(b)	(d)	(f)	(h)	(j)	(l)

*3 The formula given in the introduction can be expressed as $A = \dfrac{\theta}{360} \times C$, where A is the length of the arc and C is the circumference of the circle. Show that this formula can be changed to $C = \dfrac{A}{\theta} \times 360$.

*4 Use the formula given at the end of question 3 to find the circumference of the circle in which an arc 6 cm long subtends an angle of 54° at the centre.

5 Change $A = \dfrac{\theta}{360} \times C$ to make θ the subject. Hence find the angle (θ) subtended by an arc (A) 22 cm long, in a circle of radius 10.5 cm. (Note: First find C using $C = \pi d$, taking $\pi = \frac{22}{7}$.)

6 (a) Using D for diameter and A for arc, the arc formula becomes $A = \dfrac{\theta}{360} \times \pi \times D$.

Rewrite this formula to make the subject:
(i) D (ii) θ.

(b) In a circle of what diameter does an arc of length 15.5 cm subtend an angle of 200° at the centre? (Take $\pi = 3.1$.)

(c) What angle is subtended at the centre of a 3.5 cm radius circle by a 5.5 cm arc? (Take $\pi = 3\frac{1}{7}$.)

7 When the length of arc equals the radius of the circle, the angle subtended at the centre is 1 radian (1^c).

In Figure 7:4, arc $AB = R$, $\theta = 1^c$, and the circumference $= 2 \times \pi \times R$. Using the arc formula in the introduction this gives

$R = \dfrac{1^c}{360°} \times 2 \times \pi \times R \rightarrow R = \dfrac{1^c}{180°} \times \pi \times R \rightarrow$

$R = \dfrac{\pi^c}{180°} \times R$

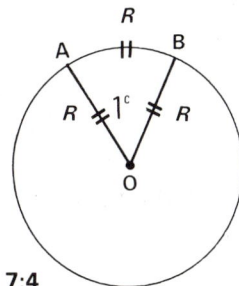

Fig. 7:4

Hence **π radians = 180°**. This important fact should be learnt.

(a) Use a calculator to find 1 radian in degrees.

(b) Show that when angle θ is measured in radians, the arc formula becomes **arc = θ × radius**. Hence find the length of arc subtending 3^c in a circle of radius 8 cm.

8 Computers usually use radian measure for angle calculations. Write a BASIC program line to convert an angle from degrees (D) to radians (R).

9 In Figure 7:5, h is the height of the cone, and l is the slant height. Figure 7:6 shows the net of the cone.

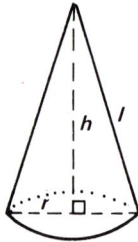

Fig. 7:5

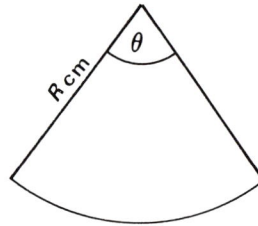

Fig. 7:6

Calculate R in cm correct to 3 s.f. and θ in radians and in degrees (to the nearest 1°) if the cone is to be 8 cm high with base radius 5 cm. (Take $\pi = 3.14$.)

Make the cone to check your answer.

B Sectors

In Figure 7:7, P is the major sector and Q is the minor sector.

Area of Q $= \dfrac{\theta°}{360°} \times$ area of circle

$\qquad\quad = \dfrac{\theta}{360} \times \pi \times r^2.$

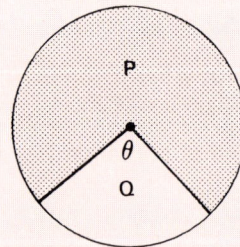

Fig. 7:7

1 For each part of Figure 7:3 find the area of the sector containing the thicker arc.

*2 Calculate the total perimeter and the area of a sector which subtends an angle of 200° at the centre of a 10 cm radius circle. (Take $\pi = 3.1$.)

3 Show that if S represents the area of the sector, then the formula given in the introduction can be changed to

$r = \sqrt{\dfrac{S \times 360}{\pi \times \theta}}.$

4 What is the radius of the circle in which a sector of area 46.5 cm^2 subtends 54° at the centre? (Take $\pi = 3.1$.)

5   (a) Remembering that π radians = 180° (see Exercise 7A, question 7), show that if θ is measured in radians the formula given in the introduction becomes

$$\text{sector area} \;=\; \frac{\theta R^2}{2}.$$

(b) What is the area of the sector subtending 2.5^c at the centre of a 3 cm radius circle?

(c) What is the angle in radians subtended at the centre of a 4 cm radius circle by a sector of area 50 cm²?

---

6   Calculate the total surface area of a cone of height 10 cm and base radius 3 cm. (Take π = 3.14.)

7   Prove that the curved surface area of a cone is $\pi r l$, where l is the slant height and r is the base radius.

8   Write a computer program to calculate the length of an arc and the sector area, given the radius of the circle and the angle in degrees subtended by the arc at the centre.

●   You need to know . . .

● Pythagoras' Theorem

In all right-angled triangles, the square on the hypotenuse is equal to the sum of the squares on the other two sides.

In Figure 8:1, $h^2 = a^2 + b^2$

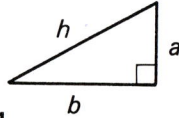

Fig. 8:1

The $\boxed{x^2}$ key is useful for finding the squares.

Examples   In Figure 8:2
$$h^2 = 5.6^2 + 8.2^2$$
$$h^2 = 98.60$$
$$h = \sqrt{98.60} \simeq 9.9\,\text{cm}$$
Key: 5.6 $\boxed{x^2}$ $\boxed{+}$ 8.2 $\boxed{x^2}$ $\boxed{=}$ $\boxed{\sqrt{}}$

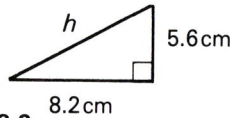

Fig. 8:2

In Figure 8:3
$$9.5^2 = 5.7^2 + a^2$$
$$a^2 = 9.5^2 - 5.7^2$$
$$a^2 = 57.76$$
$$a = \sqrt{57.76} \simeq 7.6\,\text{cm}$$
Key: 9.5 $\boxed{x^2}$ $\boxed{-}$ 5.7 $\boxed{x^2}$ $\boxed{=}$ $\boxed{\sqrt{}}$

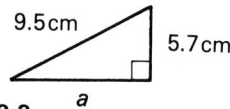

Fig. 8:3

Test yourself

1   Calculate the unknown side in each triangle in Figure 8:4. Give your answers correct to 1 decimal place.

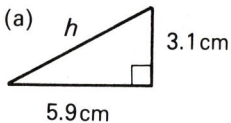

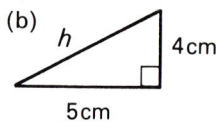

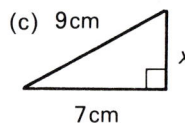

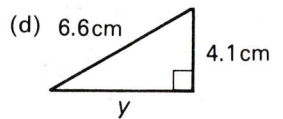

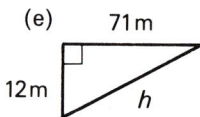

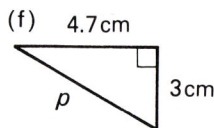

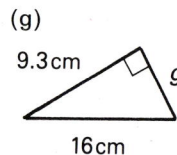

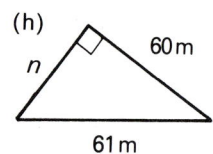

(a) h 3.1 cm   5.9 cm

(b) h 4 cm   5 cm

(c) 9 cm x 7 cm

(d) 6.6 cm   4.1 cm y

(e) 71 m   12 m h

(f) 4.7 cm p 3 cm

(g) 9.3 cm g 16 cm

(h) n 60 m   61 m

Fig. 8:4

2   Investigate the extension of Pythagoras' Theorem to acute- and obtuse-angled triangles.

A Tangent ratio: finding a side

For Discussion

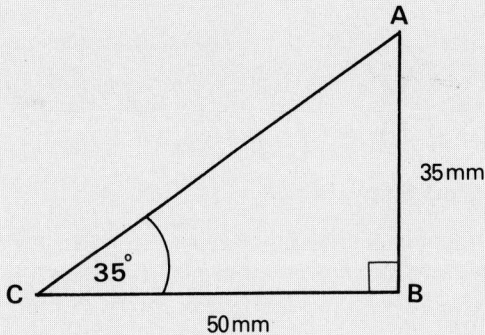

Fig. 8:5

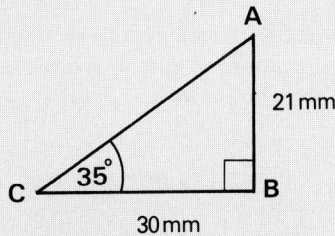

Fig. 8:6

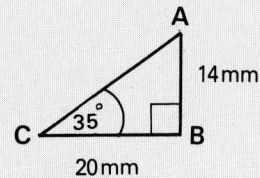

Fig. 8:7

(a) What is the same and what is different about the three triangles drawn full size in Figures 8:5, 8:6 and 8:7?

(b) For each triangle calculate AB ÷ BC.

(c) Using a scientific calculator, enter 35 $\boxed{\text{TAN}}$.

To Be Learnt

In the right-angled triangle drawn in Figure 8:8,

o is the side *o*pposite the angle θ

a is the side *a*djacent (next to) the angle θ

h is the *h*ypotenuse

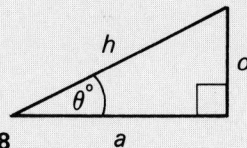

Fig. 8:8

(Note that both *a* and *h* are adjacent to θ, so more strictly we should refer to *a* as 'adjacent, not the hypotenuse'.)

o ÷ *a* is called the **tangent** of the angle θ

$$\frac{o}{a} = \tan \theta$$

Changing the subject to *o* gives

$$o = a \times \tan \theta$$

You must learn both of these formulae. It may help you to note that in both of them the order of writing the letters is *o a t.*

Using the TAN key

A scientific calculator will have a TAN key, which calculates the value of $o \div a$ for any given angle.

Look at Figure 8:9. Remembering that $\dfrac{o}{a} = \tan \theta$, what should the value of tan 45° be? Test your answer with your calculator. (Make sure that you key the 45 before the TAN.)

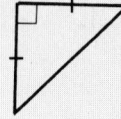

Fig. 8:9

Using tangent to find the side of a right-angled triangle

A scientist wants to measure the height of a new volcanic island. He knows from a satellite photograph that the nearest point of land is 1.25 km from the centre of the island. Taking his theodolite to this point he measures the angle between the horizontal and a line to the top of the volcano as 19°.

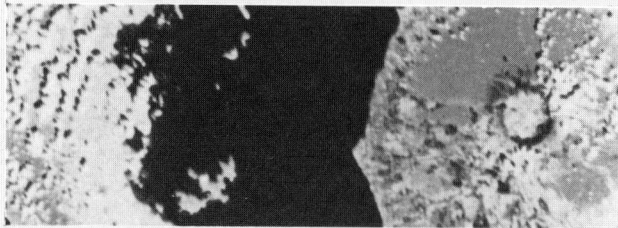

(Nigel Press Associates)

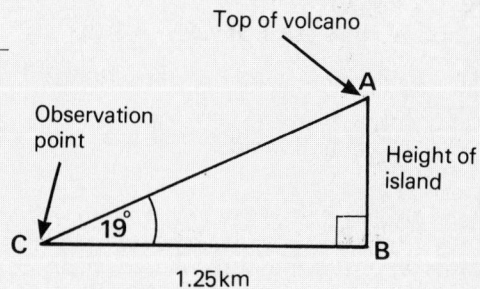

Fig. 8:12

From his simplified sketch, the scientist knows that BC is 1250 m and θ is 19°. Using $o = a \times \tan \theta$, he knows that AB = BC $\times$ tan θ = 1250 $\times$ tan 19°.

Keying on his calculator 1250 $\boxed{\times}$ 19 $\boxed{\text{TAN}}$ $\boxed{=}$ he finds that the volcanic island is about 430 metres high.

Then it starts to erupt again!!

1   **Example**   To calculate HT in Figure 8:13.

          HT = 7.6 $\times$ tan 57°

          Key: 7.6 $\boxed{\times}$ 57 $\boxed{\text{TAN}}$ $\boxed{=}$

          Answer: HT $\simeq$ 11.7 cm to 3 s.f.

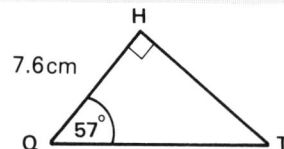

Fig. 8:13

Calculate, correct to 3 s.f., the side marked x in each triangle in Figure 8:14.

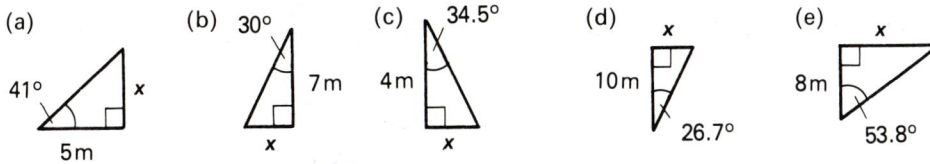

Fig. 8:14

*2 In ΔABC, AB is opposite ∠C and AC is adjacent to ∠C, so AB = AC × tan C.
Similarly, AC is opposite ∠B and AB is adjacent to ∠B, so AC = AB × tan B.
Note: BC is the hypotenuse. We do not use the hypotenuse with tangents.

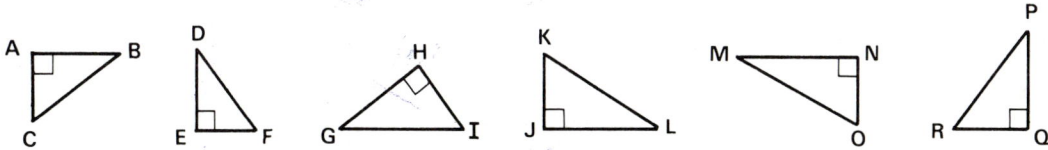

Fig. 8:15

Copy and complete the following for the other triangles in Figure 8:15.
(a) DE = EF × tan . . .   (b) EF = . . . × tan D   (c) HG = . . . × tan I
(d) HI = HG × tan . . .   (e) KJ = JL × . . .   (f) JL = . . . × tan . . .
(g) MN = . . . × . . .   (h) NO = . . . × . . .   (i) PQ = ...   (j) RQ = ...

*3 **Example**   In Figure 8:16
$$a = b \times \tan \alpha \text{ and } b = a \times \tan \theta$$

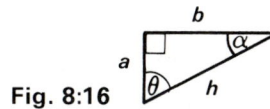

Fig. 8:16

Write similar formulae for sides c, d, e, f, g, k, l and m in Figure 8:17.

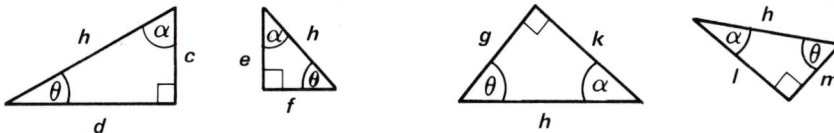

Fig. 8:17

4 For the three triangles in Figure 8:18 use tangents to calculate x correct to 3 s.f., then Pythagoras' Theorem to find h correct to 3 s.f.

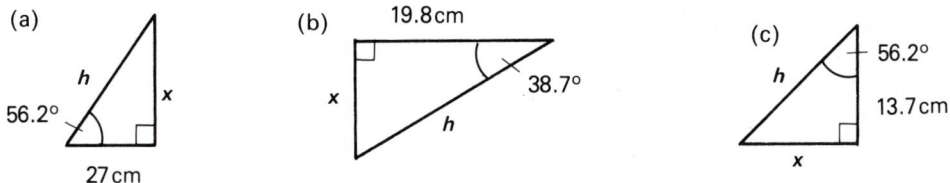

Fig. 8:18

5 Degrees, minutes, and calculators

Degrees (°) are often subdivided into minutes (′), where 60′ = 1°. Some calculators have special keys to convert between degrees/minutes and decimals of a degree. If yours has this function, learn how to use it. Make up a simple example, e.g. 10° 30′ ↔ 10.5°.

If you do not have a special key, to change an angle in degrees and minutes to decimals key in the angle as a mixed number.

Example $10° 30′ \to 10\frac{30}{60} \to$ 10 $\boxed{+}$ 30 $\boxed{÷}$ 60 $\boxed{=}$

The reverse process is a little more complex. Write down the whole degrees, then subtract this from the total and multiply the remainder by 60 to obtain the minutes.

Example $10.5° \xrightarrow{\text{write } 10}$ $\boxed{-}$ 10 $\boxed{=}$ $\boxed{×}$ 60 $\boxed{=}$

By changing the angles to decimals of a degree, find the following tangents correct to 4 s.f.
(a) tan 42° 15′   (b) tan 54° 12′   (c) tan 18° 45′   (d) tan 25° 25′   (e) tan 80° 10′

6   Example   Calculate x in Figure 8:19.

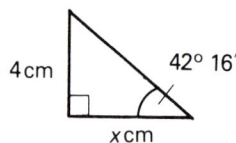

Fig. 8:19

The tangent formula calculates the length of the side **opposite** the angle whose tangent is to be found. In Figure 8:19 the side x is *not* opposite the given angle. Therefore we must work out the other angle first. This is easily done by subtracting the given angle from 90°.

Key: 90 $\boxed{-}$ 42.16 $\boxed{o'''→}$ $\boxed{=}$, which displays 47.7$\dot{3}$.

Now $x = 4 \times 47.7\dot{3}°$ etc.

The whole calculation may be done in one sequence of key presses:
90 $\boxed{-}$ 42.16 $\boxed{o'''→}$ $\boxed{=}$ $\boxed{\text{TAN}}$ $\boxed{×}$ 4 $\boxed{=}$

If you have to use the $42\frac{16}{60}°$ method to convert to a decimal of a degree you must either use brackets or change 42°16′ to 47.7$\dot{3}$° first, save it in memory, then start with the 90.

Key: 90 $\boxed{-}$ $\boxed{(}$ 42 $\boxed{+}$ 16 $\boxed{÷}$ 60 $\boxed{)}$ $\boxed{=}$ $\boxed{\text{TAN}}$ $\boxed{×}$ 4 $\boxed{=}$
Or: 42 $\boxed{+}$ 16 $\boxed{÷}$ 60 $\boxed{=}$ $\boxed{\text{MS}}$ 90 $\boxed{-}$ $\boxed{\text{MR}}$ $\boxed{=}$ $\boxed{\text{TAN}}$ $\boxed{×}$ 4 $\boxed{=}$

Calculate the side marked x, correct to 3 s.f., in each triangle in Figure 8:20.

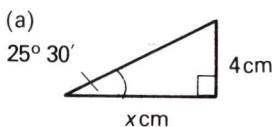

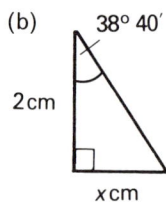

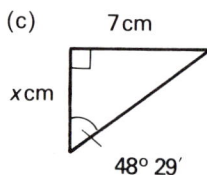

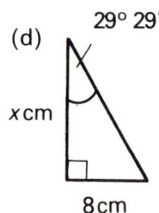

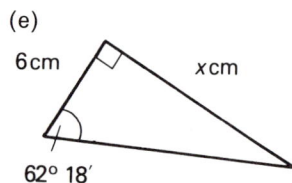

Fig. 8:20

7 A ladder reaches 6.5 metres up a wall when leaning at an angle of 80° with the ground. How far away from the wall is the foot of the ladder?

8 Two ships set sail at the same time. One sails 8 nautical miles (n.m.) due north then drops anchor. The other takes a bearing of 035° until it is due east of the first.

    (a) How far are the ships then apart?
    (b) What distance has the second ship sailed?

9 Develop a computer program that will calculate a side of a triangle given an angle and another side. Remember that computers use radians, not degrees, and that π radians = 180°.

B Tangent ratio: finding an angle

The fact that $\frac{o}{a} = \tan\theta$ can be used to calculate θ given the two sides of the triangle that include (surround) the right angle.

Key: 45 TAN . The tangent of 45° is 1.

Now we need to be able to reverse the process, so that keying in 1 will give us the 45° again. Different calculators use different methods; the most common are:

1 ARC TAN ; 1 INV TAN ; 1 ARCTAN ; 1 TAN⁻¹

We shall use ARCTAN in this book. You must key in as your own calculator requires.

Example To calculate the size of angle θ in Figure 8:21.

5.6 cm is opposite the angle θ, so $o = 5.6$
4.8 cm is adjacent to the angle θ, so $a = 4.8$

$$\frac{o}{a} = \tan\theta \rightarrow \frac{5.6}{4.8} = \tan\theta$$

Key: 5.6 ÷ 4.8 = ARCTAN

Answer: θ is 49.4° to the nearest tenth of a degree.

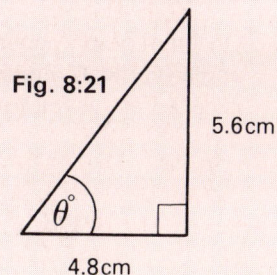

Fig. 8:21
5.6 cm
4.8 cm

1 Calculate the angle θ, correct to the nearest 0.1°, in each of the triangles in Figure 8:22. (Be careful that you use the correct two sides. We have marked the hypotenuse as well in some!)

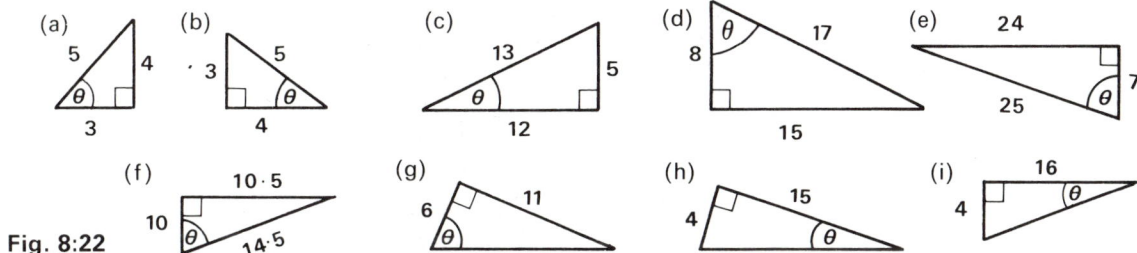

(a) 5, 4, 3
(b) 5, 3, 4
(c) 13, 5, 12
(d) 8, 17, 15
(e) 24, 25, 7
(f) 10·5, 10, 14·5
(g) 6, 11
(h) 4, 15
(i) 16, 4

Fig. 8:22

*2 **Example**   If $\tan \theta = \frac{4}{7}$ then we could sketch any of the triangles in Figure 8:23.

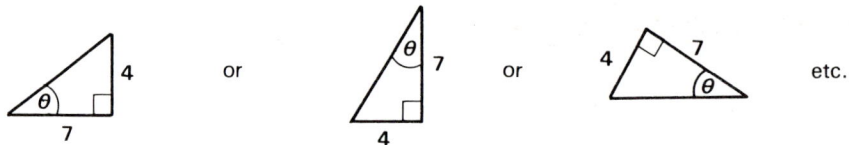

Fig. 8:23

Sketch three triangles with their right angles in different positions for each of the following.
(a) $\tan \theta = \frac{8}{11}$ (b) $\tan \theta = \frac{12}{7}$ (c) $\tan \theta = \frac{25}{6}$

3   For each of the following sketch $\triangle ABC$ with $\angle B$ a right angle. Then mark on your sketch the given lengths and calculate the required angle, correct to the nearest $0.1°$.
(a) $AB = 2.4$ cm; $BC = 1.6$ cm; calculate $\angle C$.
(b) $AB = 3.9$ cm; $BC = 26$ cm; calculate $\angle C$.
(c) $AB = 2.8$ cm; $BC = 6.3$ cm; calculate $\angle A$.

4   Sketch each of the following triangles PQR, with $\angle R = 90°$, then calculate both $\angle P$ and $\angle Q$, correct to the nearest $0.1°$.
(a) $PR = 7.6$ m; $RQ = 5$ m   (b) $PR = 18$ m; $RQ = 15$ m
(c) $PR = 57$ m; $RQ = 27$ m   (d) $PR = 3.2$ m; $RQ = 8.8$ m

---

5   Although it is not possible to have an angle bigger than 90° in a right-angled triangle, the tangent ratio continues to have values up to 360° (and beyond, although it then starts to repeat the cycle). Most calculators will give you the value of the tangent up to 360°.

Draw up a table of the tangents of every angle from 0° to 360° at 5° intervals, giving each correct to 2 s.f. (Note: tan 90° and tan 270° are infinitely large.)

On 1 mm graph paper draw a graph of Angle in degrees (horizontal) against Tangent of angle (vertical). Figure 8:24 shows you how to draw the curve at 90° and 270°. Line AB is called an 'asymptote'. The curve gets closer and closer to this line but never touches it.

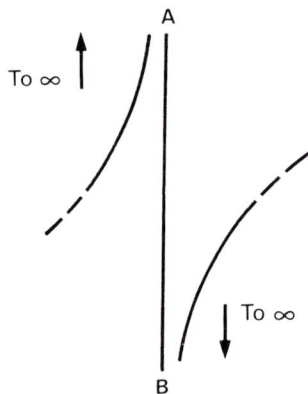

Fig. 8:24

6   By sketching suitable right-angled triangles, explain why tan 0° = 0, tan 45° = 1, and tan 90° = ∞ (infinity).

9 Ratios: volumes of similar shapes

● You need to know . . .

● The meaning of a ratio

A ratio states the connection between two quantities, e.g. the ratio of cheese to eggs for a cheese omelette could be 20 g cheese to every one egg.

Ratios are often expressed with a colon (:), e.g. 2 : 3 (say '2 to 3'). In this case, both numbers must be in the same units, so we could not write the omelette example in this way, but we could say that the ratio of the weights of flour to margarine for plain scones is 4 : 1. That is, you use 4 times as much flour as margarine, e.g. 400 g flour and 100 g margarine.

● Simplifying ratios

Ratios may be simplified by dividing by a common factor, as we do with fractions.

Example 2 litres water to 12 cl Jeyes Fluid

$\xrightarrow{\text{becomes}}$ 200 : 12 (both units are now cl)

$\longrightarrow$ 50 : 3 (dividing both 200 and 12 by 4).

● Expressing a ratio in the form $n : 1$ and $1 : n$

Ratios are easier to use if one of the quantities is 1, e.g. a ratio $1 : 2\frac{1}{4}$ clearly shows that the second amount is $2\frac{1}{4}$ times the first; this is not so obvious when the same ratio is written as 4 : 9.

Example Express 17 : 6 in the ratio: (a) $n : 1$ (b) $1 : n$.

(a) $17 : 6 \xrightarrow{\div \text{both by 6}} 2\frac{5}{6} : 1$

(b) $17 : 6 \xrightarrow{\div \text{both by 17}} 1 : \frac{6}{17}$

● Given one amount, to find the other

Example A 200 g jar of coffee granules makes about 120 cups of coffee.
The ratio of coffee to cups is 200 to 120 → 5 to 3.
Therefore a 250 g jar should make about $250 \times \dfrac{3}{5}$ cups = 150 cups.

For 50 cups we need about $50 \times \dfrac{5}{3} \simeq 85$ g of coffee.

Note how the ratio 5 : 3 became $\frac{3}{5}$ or $\frac{5}{3}$, depending on whether the required answer is to be bigger or smaller than the given amount.

● Given the total, to find each (divide in a ratio)

Example Concrete for a path should consist of 1 part cement, 2 parts sand, and 3 parts coarse aggregate. If 3 m³ of dry mix is required, what volume of each material should be purchased?

Cement: sand: aggregate = $1:2:3$
Total = $1 + 2 + 3 = 6$ parts
$3\,m^3$ in 6 parts → $\frac{1}{2}\,m^3$ per part
∴ use $1 \times \frac{1}{2} = \frac{1}{2}\,m^3$ cement;
$2 \times \frac{1}{2} = 1\,m^3$ sand;
$3 \times \frac{1}{2} = 1\frac{1}{2}\,m^3$ aggregate.

● Changing in a ratio

Example   Increase 16 in the ratio $11:6$.
An increase, so multiply by $\frac{11}{6}$.
$$\frac{{}^8\cancel{16} \times 11}{\cancel{6}_3} = \frac{88}{3} = 29\frac{1}{3}$$

Example   Decrease 35 in the ratio $6:11$.
A decrease, so multiply by $\frac{6}{11}$.
$$\frac{35 \times 6}{11} = \frac{210}{11} = 19\frac{1}{11}$$

Example   6 men can paint a school in 46 days. How long should 8 men take?
The number of men has increased in the ratio $8:6$.
The time taken should *decrease* in the ratio $6:8$.
$$\frac{{}^{23}\cancel{46} \times \cancel{6}^3}{\cancel{8}\,{}_{\cancel{4}2}} = \frac{69}{2} = 34\frac{1}{2} \text{ days}$$

● Using a conversion graph

When two quantities are in a constant ratio, a straight-line conversion graph may be used to convert one to the other. Figure 9:1 shows conversion from gallons to litres.

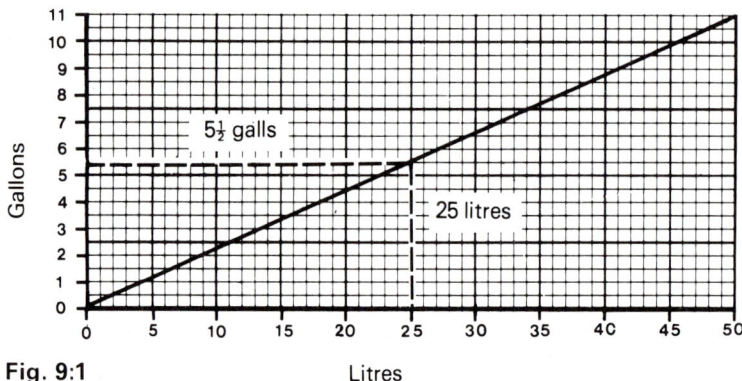

Fig. 9:1

Test yourself

1   Write the following ratios (i) in the form $x:y$ as simply as possible, (ii) in the form $1:n$, (iii) in the form $n:1$.
   (a) $21:30$ (b) 5 kg to 200 g   (c) 2 cm^2 to 3 mm^2 (d) 5 m to 3 km

2 Increase £18 in the ratio 6:5.

3 Decrease 400 g in the ratio 4:5.

4 Share £360 in the ratio 5:7.

5 *Recipe for Nettle Soup*

400 g nettle tops   30 g butter
1 large onion $\frac{3}{4}$ litre stock
15 cl cream
Spinach, salt, pepper, lemon juice, to taste.
Cook for 30 min.
Serves 5

Alter the recipe for a dinner party of 8 guests.

6 The carat rating of any metal containing gold is found by dividing its weight into 24 equal parts. A 20-carat ring would be 20 parts pure gold and 4 parts other metal.

(a) Write the ratio of gold to other metal in a 16-carat ring.

(b) The ring has no stone and weighs 4.5 g. What weight of gold is in it?

7 A theatre offers one free ticket for every twelve purchased in a block-booking. The seats cost £5 each. How much should a party of 40 pay each if they are to share the cost equally?

8 The Wunderland Army, Air Force and Navy are given their share in the annual defence budget in the ratio 3:5:2 respectively.

(a) How much should each service receive if the 1986 budget is 3 million Wunderland dollars?

(b) If the army received 195 000 dollars in 1984 how much was the whole budget?

9 A Headmaster allows his Head of Mathematics £3500 a year to supply books and materials to the 1250 pupils in the school. At the same rate, how much should he allow the German Department, which teaches only 450 pupils?

10 Investigate the ratio of weight supportable to length of span for the model bridge illustrated in Figure 9:2.

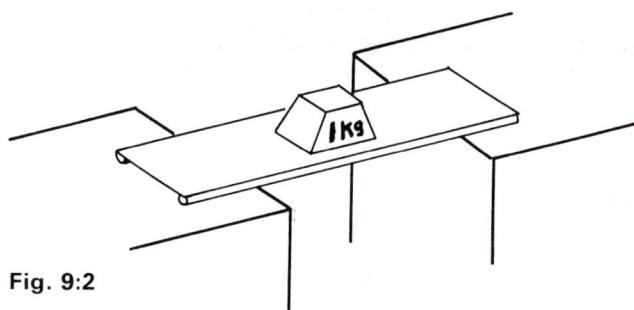

Possible components

Paper and straws

Thin card

Two paper boxes

Paper box with folded paper insert

Fig. 9:2

Go on to investigate other bridge designs.

11 Design a conversion graph to enable a painter to find how many litres of paint he needs to paint different areas of wall.

12 $x:y = 3:4$ and $y:z = 6:7$. Write as simply as possible the ratio $x:y:z$.

13 Investigate gear ratios, e.g. for a car, a cycle, or a lathe.

A   Lengths and areas of similar shapes: review

For Discussion
The three photographs and their frames shown in Figure 9:3 are mathematically similar shapes. The smallest uses 24 cm of moulding and contains 32 cm² of glass. Discuss the lengths of moulding and the areas of glass for the other frames.

5 cm

12 cm

(London Mathematical Society)

10 cm

Fig. 9:3

The following program illustrates that shapes are only similar if their lengths are all multiplied by the same factor. Adding the same amount ('the addition strategy') does not give similar shapes.

Consider also how the areas increase as the lengths change.

Main program
```
10 REM "SIMDEM"
20 CLS (Clear screen)
30 PRINT "The following rectangles are
   similar. All lengths are multiplied by the
   same factor."
40 LET S = 1
50 GOSUB 130
60 PRINT "Press RETURN to continue."
   (ENTER on Sinclair)
70 INPUT A$
80 CLS (Clear screen)
90 PRINT "The following rectangles are not
   similar. Lengths increase by adding 1
   each time."
100 LET S = 0
110 GOSUB 130
120 END
```

Subroutine to draw rectangles
```
130 LET C = 1
140 LET E = 2
150 LET F = 1
160 LET G = 2
170 FOR A = 1 TO 4
180 FOR B = 1 TO C
190 FOR D = 1 TO E
200 LET P$ = "*"
210 IF B > F OR D > G THEN LET
    P$ = "+"
220 PRINT P$;
230 NEXT D
240 PRINT
250 NEXT B
260 LET F = C
270 LET G = E
280 IF S = 1 THEN LET C = C*2
290 If S = 0 THEN LET C =
    C + 1
300 IF S = 1 THEN LET E = E*2
310 IF S = 0 THEN LET E = E + 1
320 PRINT
330 NEXT A
340 RETURN
```

Note: (a) The use of the 'flag' variable, S, to set the subroutine to draw the two different sets of rectangles.

(b) The 'nested loops' in lines 170 to 330. One loop must be completely enclosed by any other loop.

1 The pet food manufacturer Petco promises that for every three Petco can-labels sent to them it will donate 2p to the R.S.P.C.A.

(a) How much will Petco donate if I send them 30 labels?

(b) How many labels must I send for a donation of £1?

2 Two different scale models of a railway engine are available. One model fits a 20 mm wide track; the other fits a 25 mm wide track. Show that the models' lengths are in the ratio 4 : 5, then calculate the length of the longer engine if the smaller is 14 cm long.

3 Two squares have sides of 2 cm and 5 cm. Write as simply as possible the ratio of:
(a) their perimeters   (b) their diagonals   (c) their areas.

4 Draw two similar rectangles with sides in the ratio 2 : 3, making the smaller one 2 cm long and 1 cm wide. By drawing, or otherwise, find the ratio of their areas.

5 Example The two triangles in Figure 9:4 are similar.

Looking at the positions of the sides we can see that 4 cm $\xrightarrow{\text{becomes}}$ 6 cm, a cm → 5 cm, and 2 cm → b cm.

Therefore their sides are in the ratio 4 : 6 = 2 : 3.

a is smaller than 5, so $a = \dfrac{2}{3} \times 5 = 3\frac{1}{3}$ cm.

b is larger than 2, so $b = \dfrac{3}{2} \times 2 = 3$ cm.

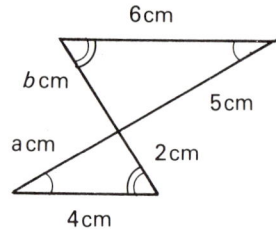

Fig. 9:4

Note that the areas of the triangles are in the ratio $2^2 : 3^2 = 4 : 9$. That is, the larger triangle has sides only half as big again as the smaller, but its area is $2\frac{1}{4}$ times as big.

Remember: Areas' ratio = square of lengths' ratio

In Figure 9:5 calculate a to g for the pairs of similar shapes. All lengths are in centimetres.

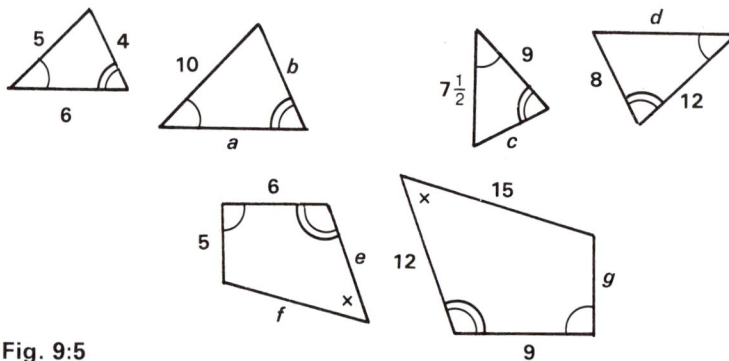

Fig. 9:5

6 State, as simply as possible, the areas' ratio for each pair of similar shapes in Figure 9:5. Also state how many times bigger in area the larger one is compared to the smaller one.

7 In Figure 9:6 the pairs of shapes are similar. Calculate areas (a), (b), and (c).

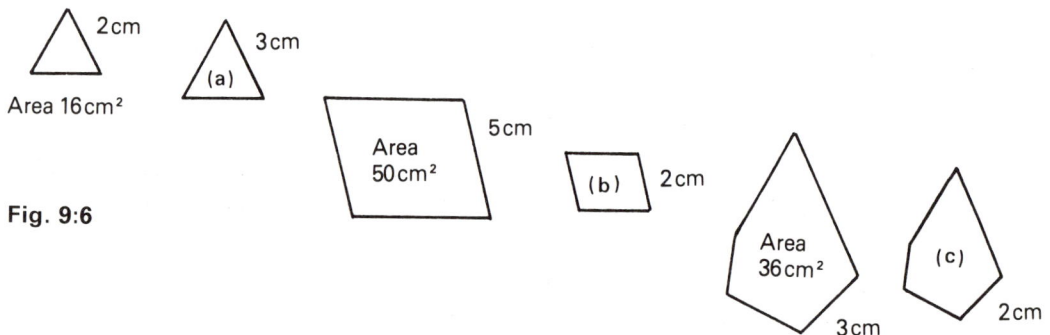

Fig. 9:6

8 The three tiers of a wedding cake are to be similar shapes with sides in the ratio 5:4:3. The largest tier is 30 cm square and 10 cm deep.

 (a) Calculate the sizes of the other two tiers.

 (b) Each cake is covered with the same thickness of marzipan and icing. The bottom tier needs 250 g of marzipan and 100 g of icing. How much of each is needed for the other tiers?

*9 The plans of an office block are made to a scale of 1:50. One room is shown to be 13.4 cm wide and 15.2 cm long on the plan. What will be its true size, in metres?

*10 Express these map ratios in their simplest form.

 Example 3 cm represent 450 m → 3 cm represents 45 000 cm → 3:45 000 → 1:15 000

 (a) 1 cm represents 500 m (b) 4 cm represents 600 m (c) 5 cm represents 1 km

*11 A map is drawn to a scale 1:100 000.

 (a) What length in km does 1 cm on the map represent?

 (b) A forest measures 1 cm by 2.4 cm on the map. What are its real measurements in km?

*12 For the pairs of similar shapes in Figure 9:7, state the areas (a), (b) and (c).

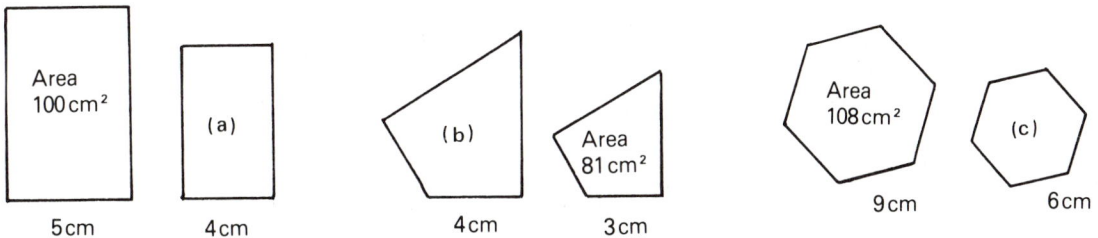

Fig. 9:7

13 A photograph is 9 cm long and has an area of 54 cm². What is the area of an enlargement of the photograph which is 12 cm long?

14 A circle has an area of 12 cm². What is the area of a circle with a radius:
 (a) twice as big (b) three times as big (c) half as big?
 (Note: Do not use the area of a circle formula.)

15 Two similar polygons have areas of 810 cm² and 160 cm² respectively. If one side of the larger polygon is 9 cm long, find the length of the corresponding side on the smaller polygon.

---

16 For maps and plans the ratio is usually in the form 1 : *n*, called the Representative Fraction (R.F.).

 Example The R.F. of a map is 1 : 60 000.
   This means 1 cm represents 60 000 cm = 600 m.

Find the actual distances represented by the given distances on a map with the given R.F.:

(a) 3 cm, R.F. = 1 : 1000   (b) 5 cm, R.F. = 1 : 10 000

(c) 9 cm, R.F. = 1 : 20 000   (d) 7 cm, R.F. = 1 : 100 000

(e) 7 mm, R.F. = 1 : 25 000   (f) 4.6 cm, R.F. = 1 : 50 000.

17 An Ordance Survey map, R.F. 1 : 50 000, measures 80 cm by 80 cm.

(a) What ground distances do these measurements represent?

(b) The map is subdivided into squares of 2 cm sides and of 20 cm sides. What distances do these measurements represent?

(c) What ground area is represented by: (i) the map   (ii) the small squares   (iii) the large squares?

18 Find the real areas, in km^2, represented by the following areas on maps with the given R.F.

(a) 25 cm^2; R.F. 1 : 10 000   (b) 46 cm^2; R.F. 1 : 20 000

19 Investigate the ratios of height to weight for men and for women. Report on your findings, including charts and graphs.

20 The following program illustrates the way that lengths and areas increase as a square grows. Lines 30 to 260 are used twice. The first time, the squares are drawn; the second time the resulting changes are shown in a table.

```
10 REM "SQUARE"
20 LET R = 0
30 FOR S = 1 TO 19
40 IF R = 1 AND S > 1 THEN GOTO 70
50 CLS (Clear screen)
60 PRINT "Side"; TAB (11); "Perimeter"; TAB
   (24); "Area"
70 PRINT; S; TAB (3); "(+1)"; TAB (11); 4*S;
   TAB (15); "(+4)";
80 IF S < 2 THEN GOTO 110
90 LET D = S*S
100 PRINT TAB (24); D; TAB (29); "(+ ^"; D − C;")"
110 LET C = S*S
120 PRINT
130 IF R = 1 THEN GOTO 260
140 FOR A = 1 TO S + 7
150 LET X = 1
160 LET Y = A
170 PLOT X, Y + 5 (Add ,2 on a 380/480Z )
180 IF Y = 1 THEN GOTO 220
190 IF X = A THEN LET Y = Y − 1
200 IF X < A THEN LET X = X + 1
210 GOTO 170
220 NEXT A
230 PLOT 0, 0 (You may not need this line)
```

Changes for BBC

```
5 MODE 4
6 PRINT CHR$ (14)
170 PLOT 69, X*50,Y*50
```

Notes

(a) Flag R sets the program to either draw the squares or show the table.

(b) Variable D is set to the area of the previous square; variable C is the area of the current square.

(c) Loop 140 to 220 draws the square.

```
240 PRINT "Press RETURN to continue."
      (ENTER on Sinclair)
250 INPUT C$
260 NEXT S
270 IF R = 1 THEN END
280 LET R = 1
290 GOTO 30
```

B Ratio of volumes

The ratio of the volumes of similar solids is the cube of the ratio of their corresponding lengths.

For Discussion

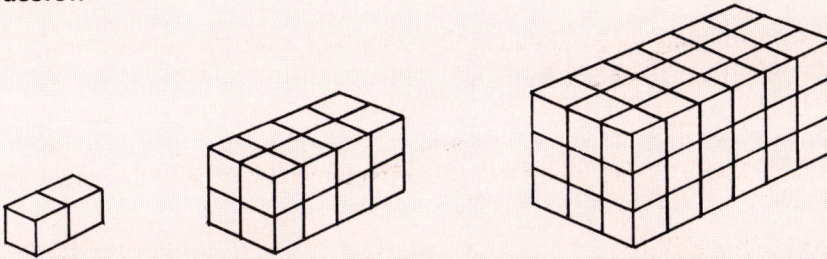

Fig. 9:8

Examples (a) A vase, similar in shape to the one in Figure 9:9 but twice as tall, holds 2 litres. Find the dimensions of this vase, and the capacity of the vase drawn in Figure 9:9.

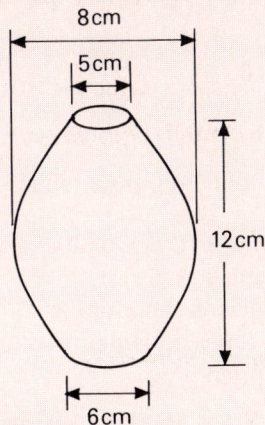

8 cm
5 cm
12 cm
6 cm

Fig. 9:9

(b) Lead shot is made as spheres of various radii. A 1 mm radius sphere weighs 50 mg. Find the weight of other shot, up to a radius of 10 mm.

9

1   A plastic cube of side 1 cm weighs 1 g. State the weight of a cube made from the same plastic with side:
(a) 2 cm   (b) 3 cm   (c) 4 cm   (d) 10 cm.

2   Two similar jugs have heights of 3 cm and 6 cm. The smaller has a base radius of 5 cm. The larger has a volume of 480 cm³.

(a) What is the base radius of the larger jug?

(b) What is the volume of the smaller jug?

3   A block is 1 cm wide, 2 cm long and 3 cm high. State the dimensions of a similar block made from the same material, but eight times as heavy.

4   Two similar cuboids have their lengths in the ratio 3 : 5. The larger has a volume of 250 cm³.

(a) State the ratio of the volumes of the two cuboids.

(b) Calculate the volume of the smaller cuboid.

5   A sphere of lead weighs 100 g. Calculate the weight of a sphere of lead of:
(a) twice the radius   (b) three times the radius   (c) ten times the radius
(d) half the radius   (e) one tenth the radius.

*6   A packet of washing powder weighs 5 kg. What would be the weight of a similar shaped packet:
(a) twice as tall   (b) three times as tall?

*7   A marble model of a marble statue weighs 12 kg. The original statue is ten times the height of the model. What is its weight?

8   A glass marble weighs 30 g and is 10 mm in diameter.

(a) How much will eight of the marbles weigh?

(b) The eight marbles are melted down and recast into one large one. Write down its radius. (No working is necessary.)

9   Two bottles are of similar shapes. The one which is 30 cm tall holds 1 litre when full.

(a) How much does the other, 15 cm tall, hold when full?

(b) If a similar bottle is made to hold 1 ml, how tall is it?

10 **Example**   The two boxes in Figure 9:10 are similar solids.
The ratio of their lengths is 3 : 5.
The ratio of their volumes is therefore $3^3 : 5^3 = 27 : 125$.
The larger is $\frac{125}{27} = 4\frac{17}{27}$ times as large in volume, but only
$\frac{5}{3} = 1\frac{2}{3}$ times as large in lengths.

Volume of smaller $= \dfrac{27}{125} \times$ volume of larger

$= \dfrac{27}{125} \times 75 \text{ cm}^3 = 16.2 \text{ cm}^3.$

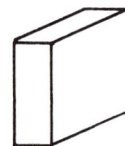
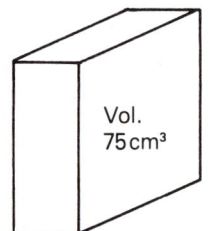

Vol. 75 cm³

Fig. 9:10   3 cm   5 cm

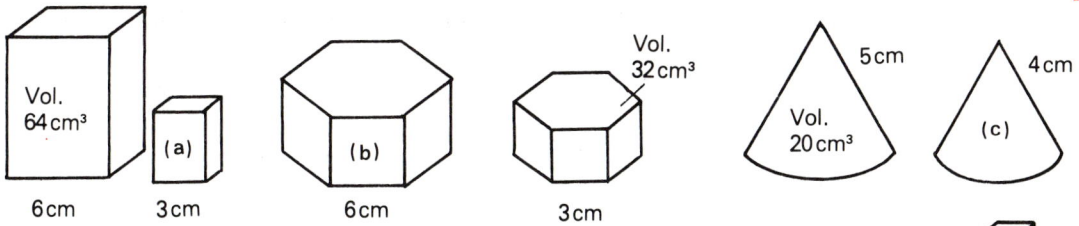

9

Vol. 64 cm³ (a)

6 cm 3 cm

(b)

Vol. 32 cm³

6 cm 3 cm

5 cm Vol. 20 cm³

4 cm (c)

Fig. 9:11

5 cm (d)

7 cm

Vol. 98 cm³

For each pair of similar solids in Figure 9:11:
(i) state the lengths' ratio
(ii) calculate the unknown volume.

11 The middle-size dwarf in Figure 9:12 weighs 5 kg. What are the weights of the others?

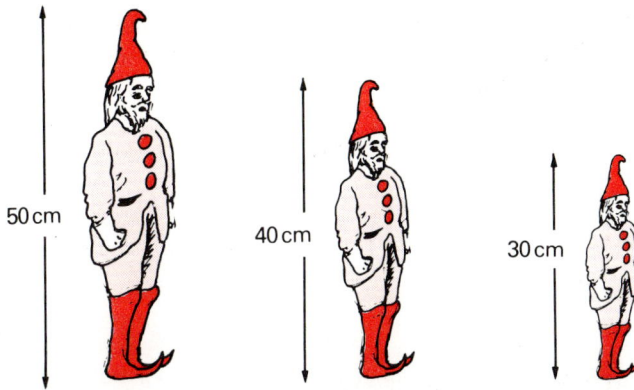

50 cm 40 cm 30 cm

Fig. 9:12

12 The fish in Figure 9:13 are all approximately similar. The smallest one provides about 250 calories. How many calories, to the nearest 10, would the other fish provide?

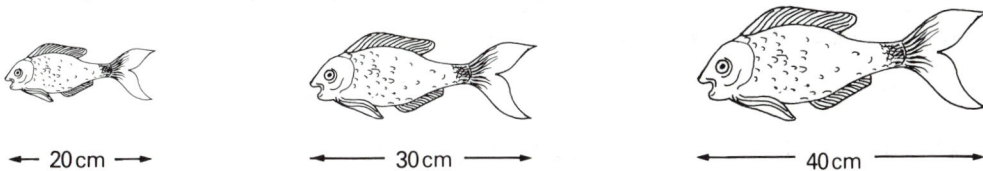

20 cm 30 cm 40 cm

Fig. 9:13

13 A sweet company markets its products in boxes of similar shapes. The 9 cm-long box holds 160 grams. What length box is needed to hold 540 grams, and how many times as much card will be needed to make it?

14 Ali can just lift 100 kg. Slabs of metal need to be lifted. They are various rectangular shapes, but all have length-to-width ratio 2 : 1, and all have the same thickness. A 30 cm by 15 cm slab weighs 10 kg. What size is the largest slab that Ali can lift?

69

15 Two similar bottles hold 1.35 litres and 3.2 litres. The larger bottle has a radius of 4 cm and uses 5p's worth of glass.

(a) Calculate the radius of the smaller bottle.

(b) Calculate the cost of the glass to make the smaller bottle, if both bottles use the same thickness of glass.

16 Kuvrup paint cans are similar shapes. Figure 9:14 shows a 1 litre can.

How high is:
(a) the $2\frac{1}{2}$ litre can   (b) the 5 litre can?

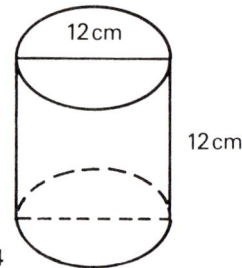

Fig. 9:14

17 A baby weighs 3.5 kg and is 50 cm long. If babies were similar 'shapes' to adults, what would be the weight of an adult, 180 cm tall, (a) in kg, and (b) in stones? (6 kg ≏ 1 stone). Discuss your answer.

18 Gog was an evil giant. He was 14 feet tall.

(a) Comparing him with a man of height 6 feet (assuming similar shapes), find:
   (i) the ratio of the heights
   (ii) the ratio of the cross-section area of the bones (the cross-section area determines the strength of bones)
   (iii) the ratio of the body volumes

(b) How many times more weight per unit area did Gog's bones have to withstand than the man's bones?

(c) Calculate the Gog's weight if an average 6-feet tall man weighs 12 stone 12 pounds (1 stone = 14 pounds).

19 The following computer program illustrates the change in the volume of a cuboid as one length, two lengths, or all three lengths (giving a similar cuboid) change.

Main program
```
10   REM "CUBE"
20   LET S = Ø: LET W = Ø
30   LET T = Ø: LET H = Ø
35   LET B = 1: LET D = Ø
40   PRINT "1 Increase height only?"
50   PRINT "2 Increase height and depth?"
60   PRINT "3 Increase all lengths in same
     ratio?"
70   PRINT "Choose 1, 2 or 3."
80   INPUT R
90   IF R = 2 OR R = 3 THEN LET T = 1
```

```
100 IF R = 3 THEN LET S = 1
110 FOR P = 1 TO 7
120 CLS (Clear screen)
130 LET H = (H + P + 2)*B
140 LET W = (3 − S + (W + P)*S)*B
150 LET D = (3 − T + (INT((D + P*T)/
    1.5)*T))*B
160 GOSUB 1000
170 IF R = 1 THEN LET V = P
180 IF R = 2 THEN LET V = P*P
190 IF R = 3 THEN LET V = P*P*P
200 PRINT "Volume = ‸"; V;"‸cubic units."
210 PRINT "Press RETURN to continue."; (Use
    ENTER on Sinclair)
220 INPUT C$
230 NEXT P
240 END
```

Subroutine
```
1000 FOR A = 1 TO H
1010 PLOT 1, A (Add ,2 for
     a 380/480Z)
1020 PLOT W, A
1030 PLOT W + D, A + D
1040 NEXT A
1050 FOR A = 1 TO W
1060 PLOT A, 1
1070 PLOT A, H
1080 PLOT A + D, H + D
1090 NEXT A
1100 FOR A = 1 TO D
1110 PLOT A + 1, A + H
1120 PLOT A + W, A + H
1130 PLOT A + W − 1, A
1140 NEXT A
1150 RETURN
```

Changes for BBC
All 'PLOT's need 69,
e.g. 1010 PLOT 69, 1, A

```
  15 MODE 4
  35 LET B = 10
 155 IF R = 1 THEN D = D/2
1145 LET H = H/B
1146 LET W = W/B
1147 LET D = D/(12 − R)
```

Notes: (a) The subroutine was not essential here, as it is only used at one point in the program, but if often helps to develop part of a program separately, as a 'module'. Here the module is concerned with the drawing of the cuboids.

(b) If you can adapt the program for high-resolution graphics the cuboids will be much more realistic.

(c) The rather complex arithmetic in lines 130 to 150 was needed to ensure a correct cuboid was drawn at all stages in the program. When S and T are 0, much of each expression reduces to zero.

● You need to know . . .

● The mathematical definition of probability

Mathematicians state the probability of an event happening as a fraction between 0 (impossible) and 1 (certain). This fraction may be written as a common or decimal fraction (e.g. $\frac{1}{4}$ or 0.25), as a ratio (1 : 4 or 1 in 4), or as a percentage (25%).

Betting odds are stated differently. Probability $\frac{1}{4}$ becomes odds of 3 to 1 (3 chances that it will not happen to 1 chance that it will).

The probability of a successful outcome (i.e. that the event you want will happen) is:

$$\frac{\text{number of successful outcomes possible}}{\text{total number of possible outcomes}}.$$

● Combining probabilities

Outcomes are **independent** when one event happening does not prevent (or exclude) the others from happening.

When the successful outcomes are independent the probabilities of each are **multiplied** to obtain the probability of them all happening.

Example   Calculate the probability of drawing first an ace and then a king from a full pack of 52 cards if:
(a) the ace is replaced after being picked
(b) the ace is not replaced.

The two successful outcomes are: (i) drawing an ace first, (ii) drawing a king second. These are independent events because after (i) happens it is still possible to go on to pick a second card to see if (ii) happens.

(a) With replacement:
$P(\text{ace first}) = \frac{1}{13}$; $P(\text{king second}) = \frac{1}{13}$
$P(\text{ace then king}) = \frac{1}{13} \times \frac{1}{13} = \underline{\underline{\frac{1}{169}}}$

(b) Without replacement:
$P(\text{ace first}) = \frac{1}{13}$; $P(\text{king second}) = \frac{4}{51}$
$P(\text{ace then king}) = \frac{1}{13} \times \frac{4}{51} = \underline{\underline{\frac{4}{663}}}$

When the successful outcomes are **exclusive** the probabilities of each are **added** to obtain the probability of one or the other happening. See the next exercise.

Test yourself

1   A tetrahedral die has four faces, marked 1, 2, 3 and 4. What is the probability of the number on the base triangle, after the die is shaken and thrown, being:
(a) 1   (b) 1, 1 (two throws)   (c) 1, 1, 1 (three throws)?

2 A bag contains 6 black, 4 white and 2 grey counters. State the probability of picking (if each picked counter is replaced):
(a) a black　　(b) a white　　(c) a black then a white　　(d) a black then a black
(e) a white then a white　　(f) a black then a grey.

3 A card is picked from a full shuffled pack, then returned before further cards are picked. Give, in as simple a form as possible, the probability of picking:
(a) a heart　　(b) a heart twice in a row　　(c) a club three times in a row
(d) any king　　(e) any king twice in a row　　(f) the King of Spades twice in a row
(g) a picture card (K, Q, or J)　　(h) an ace then a spade　　(i) a club then a heart
(j) the King of Clubs then a red card　　(k) any king, then the same king the second pick.

4 Repeat question 2, (c) to (f), if the first counter picked is not replaced (although each part starts with a full bag).

5 Repeat question 3, (b) to (f) only, if the picked cards are not replaced (although each part starts with a full pack).

6 A box contains pens of three colours: 3 red, 4 black, and 5 green. All the ways of picking three pens from the box, without replacement, can be shown on a tree diagram; see Figure 10:1, which also shows the probability of each event. Copy and complete it (fill in the '3rd pick' column first, which will have 27 rows).

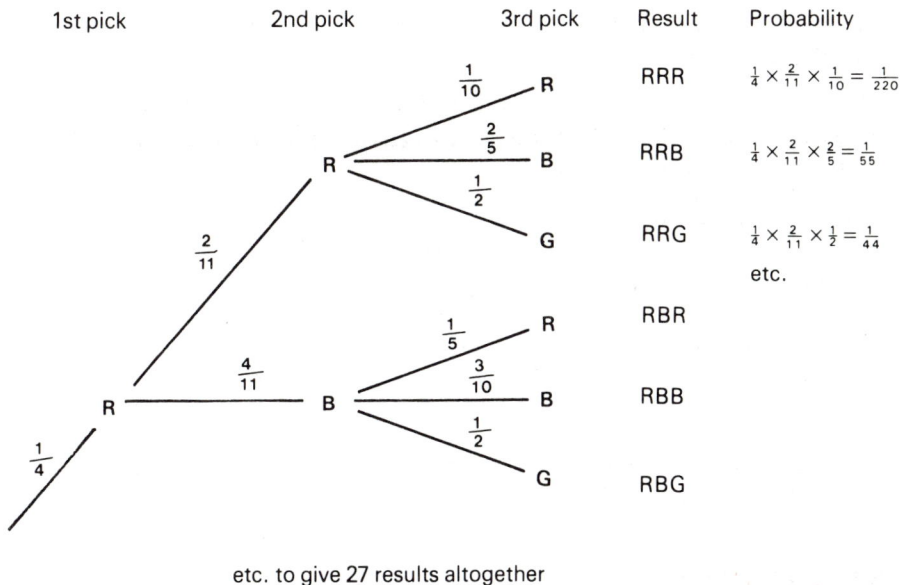

1st pick	2nd pick	3rd pick	Result	Probability
		$\frac{1}{10}$ R	RRR	$\frac{1}{4} \times \frac{2}{11} \times \frac{1}{10} = \frac{1}{220}$
	R	$\frac{2}{5}$ B	RRB	$\frac{1}{4} \times \frac{2}{11} \times \frac{2}{5} = \frac{1}{55}$
$\frac{2}{11}$		$\frac{1}{2}$ G	RRG	$\frac{1}{4} \times \frac{2}{11} \times \frac{1}{2} = \frac{1}{44}$
		$\frac{1}{5}$ R	RBR	etc.
R	$\frac{4}{11}$ B	$\frac{3}{10}$ B	RBB	
$\frac{1}{4}$		$\frac{1}{2}$ G	RBG	

etc. to give 27 results altogether

Fig. 10:1

7 Design a suitable pay-out for the outcomes (e) to (j) in question 4 if the stake is 10p a go ('Pick 2 counters.'). Test out your chosen answers experimentally to make sure that you are going to make a profit, then perhaps you could use the idea at a 'School Fayre'.

8 Computers are ideal for investigating probability as they can very rapidly produce outcomes for a large number of events. In BASIC the RND (or RND(1)) function gives a pseudo random number between 0 and 1. Many computers give the same set of random numbers each time the program is run, unless you give the instruction RAND (RANDOMIZE) at the start.

In the following program the computer simulates the random throwing of a dart at a circular target. (The equation of a circle is $x^2 + y^2 = r^2$, where r is its radius). Line 60 finds if the dart hits a circle of radius 1 unit. The formula (*4 times number of hits*) *divided by the number of throws* gives an approximation for pi.

```
 5 REM "MONTEPI"
 6 PRINT "PI by simulation of darts hitting a circular target; the Monte-Carlo
   method."
10 RANDOMIZE (May be omitted)
20 LET T = 0
30 LET H = 0
40 LET X = RND (or RND(1))
50 LET Y = RND
60 IF X*X + Y*Y < = 1 THEN LET H = H + 1
70 LET T = T + 1
80 IF T/50 = INT(T/50) THEN PRINT T;" ˄throws gives PI as ˄"; 4*H/T
90 GOTO 40
```

Exclusive events

Events are exclusive when one of them happening excludes (prevents) the others from happening.

For exclusive events, add the separate probabilities

Example Calculate the probability of picking either a king or a queen from a full pack of cards.

The successful outcomes are: (i) picking a king first pick, (ii) picking a queen first pick. Either may happen, but not both, so the two events are exclusive.
$P(\text{king}) = \frac{1}{13}$; $P(\text{queen}) = \frac{1}{13}$
$P(\text{king or queen}) = \frac{1}{13} + \frac{1}{13} = \underline{\underline{\frac{2}{13}}}$

This answer may be checked by considering that either a king or a queen gives a chance of 2 cards in every 13, or $\frac{2}{13}$.

For Discussion
(a) Picking an ace or a king.
(b) Picking an ace or a king or a queen.
(c) Scoring a 1 or a 6 with a die.
(d) Scoring a 2 or a 4 or a 6 with a die.
(e) Picking a red or a blue from 3 red, 2 blue and 5 green counters.
(f) Picking a red or a blue or a green from part (e).

1 Calculate the following probabilities.

(a) Picking from a pack:
(i) the four of spades　(ii) the five of hearts
(iii) the four of spades *or* the five of hearts.

(b) Scoring with a die:
(i) a 3　(ii) a 6　(iii) a 3 *or* a 6.

(c) Being born on:
(i) a Saturday　(b) a Sunday　(c) a Saturday *or* a Sunday.

(d) Picking from a bag of 8 black, 2 white and 2 red:
(i) a black　(ii) a white　(iii) a black *or* a white.

2 **Example**　To find the chance of exactly 2 heads in 3 tosses of a coin.

(i) List the possible ways of getting 2 heads:
HHT;　HTH;　THH.

(ii) Calculate the probabilities of each exclusive event:
$P(\text{HHT}) = \frac{1}{2} \times \frac{1}{2} \times \frac{1}{2} = \frac{1}{8};$　$P(\text{HTH}) = \frac{1}{2} \times \frac{1}{2} \times \frac{1}{2} = \frac{1}{8};$
$P(\text{THH}) = \frac{1}{2} \times \frac{1}{2} \times \frac{1}{2} = \frac{1}{8}.$

(iii) Add the exclusive probabilities: $\frac{1}{8} + \frac{1}{8} + \frac{1}{8} = \frac{3}{8}$

Notes:　(a)　You could answer the above example by listing all possible outcomes, then seeing how many of them fit the requirement of having exactly 2 heads: HHH; **HHT; HTH;** HTT; TTT; TTH; THT; **THH**.

(b)　Because of the commutative nature of multiplication the three answers in part (ii) of the example were bound to be the same, and really it was only necessary to calculate one of them.

Calculate the probability of:
(a) 1 head only in 2 tosses of a coin
(b) 2 boys only in a family of 3 (boy/girl equally likely)
(c) a jack and a queen resulting from two picks, each from a full pack
(d) 1 head only in 3 tosses of a coin
(e) a red card and an ace resulting from two picks, each from a full pack.

3 The probability of a dry day in Uptown is $\frac{1}{3}$. What is the chance of:
(a) a wet day　(b) two dry days in a row　(c) two wet days in a row
(d) one dry day, the next day wet　(e) one day wet, the next day dry
(f) one day dry out of any two days.

4 A coin is biased so that it is more likely to show 'heads': the probability of heads is $\frac{3}{5}$. What is the probability of:
(a) tails　(b) 2 heads in a row　(c) a head then a tail
(d) a tail then a head　(e) a head and a tail in either order?

*5 Calculate the probability of:
   (a) picking either the Ace of Spades or the Ace of Clubs from a full pack
   (b) scoring 5 or 6 on one throw of a die
   (c) a stranger's birthday being on a Monday *or* a Friday this year
   (d) one head and one tail in two tosses of a coin
   (e) one head in four tosses of a coin
   (f) three heads in four tosses of a coin
   (g) an ace and a king in two picks (each from a full pack)
   (h) a red ace and a black card in two picks (each from a full pack).

6 Write out all the possible exclusive events that give a successful outcome for each of the following.

   Example   A score of 6 from two dice.
                 Answer: 5, 1;   4, 2;   3, 3;   2, 4;   1, 5.

   (a) A score of 4 from two dice.
   (b) A score of 5 from two dice.
   (c) At least one head in two tosses of a coin.
   (d) At least one head in 3 tosses of a coin.

7 Calculate the probability for each part of question 6.

8 The probability that Miss Poisonfinger's pot-plants will die in a month is $\frac{5}{6}$. What is the probability of:
   (a) one of two pot-plants dying in a month
   (b) only one of three pot-plants living for more than a month?

9 A bag contains 4 pink, 3 black and 5 orange balls. By considering all the various ways each could occur, calculate the probability, with replacement of each picked ball, of:
   (a) a pink and a black (2 picks)   (b) a pink, a black and an orange (3 picks)
   (c) two pinks (2 picks)   (d) one pink and one black (3 picks)
   (e) two pinks (3 picks)   (f) no blacks (2 picks).

10 Calculate the probabilities of the following if each part is *without* replacement.

   (a) An ace then a black card in two picks. (Hint: Consider red ace/black card; black ace/black card.)

   (b) A red card then a king in two picks. (Hint: Consider (i) the red card is a king; (ii) the red card is not a king.)

   (c) A king then a hearts card in two picks. (Hint: Consider (i) the king is the King of Hearts; (ii) the king is not the King of Hearts.)

   (d) An ace and a black card in two picks. (Hint: See part (a).)

   (e) A red card and a king in two picks. (Hint: See part (b).)

   (f) A king and a hearts card in two picks. (Hint: See part (c).)

11 A fruit machine has three rotating wheels, each of which has 9 cherries, 6 lemons and 3 bells. State the probability of:
   (a) the middle wheel coming up with: (i) a bell   (ii) a lemon
   (b) all 3 wheels coming up with: (i) cherries   (ii) the same picture
   (c) all 3 wheels not coming up with the same picture (use the answer to part (b)).

12 Record in a table all the ways that three dice could give a score of 7, then state the probability that this will happen.

13 Test some of your probability answers by writing computer programs to simulate the events.

● You need to know . . .

● The metric system (SI)

Base units likely to be met in mathematics
Length: metre (m)
Mass (weight): kilogram (kg)
Time: second (s)

Prefixes in common use

mega (M) = 10^6

kilo (k) = 10^3

milli (m) = $10^{-3} \left(\dfrac{1}{1000} \right)$

centi (c) = $10^{-2} \left(\dfrac{1}{100} \right)$

micro (μ) = $10^{-6} \left(\dfrac{1}{1\,000\,000} \right)$

Other units which may be used with SI
litre (best not abbreviated) = 1000 cm^3
tonne (best not abbreviated) = 1000 kg
hectare (best not abbreviated) = 10 000 m^2

Some metric prefixes are used with these, e.g. the centilitre (cl) = $\frac{1}{100}$ litre = 10 cm^3, the millilitre (ml) = $\frac{1}{1000}$ litre = 1 cm^3, the megatonne.

Changing from one metric unit to another
Never insert zeros between figures. The figure in the units' column is rewritten in the correct column of the new unit.

Examples (a) 108.9 mm → metres
  The 8 in the units' column is 8 mm = $\frac{8}{1000}$ metre, so the 8 is rewritten in the thousandths' column, giving 108.9 mm → 0.1089 metres.

  (b) 3.06 cm → metres

  3.06 cm ────────→ 0.0306 m
  └──── 3 cm = $\frac{3}{100}$ m ────┘

  (c) 0.00306 m ──→ km

  0.00306 m ────────→ 0.000 003 06 km
  └──── 0 m = $\frac{0}{1000}$ km ────┘

● Decimal fraction arithmetic

When adding or subtracting be sure to keep the units' figures in a vertical line.

When multiplying, ignore zeros at the beginning or end of the numbers; multiply the resulting integers, then replace all omitted 'end' zeros; finally replace the decimal point so that there are as many figures after it as there were after the points in the original question.

Example   0.381 × 10 700 → 381 × 107 → 40 767 → 4 076 700 → 4 076.700 → <u>4076.7</u>

When dividing, multiply both numbers by the power of 10 needed to change the divisor (the number you are dividing by) into an integer, then divide as usual. No further change in the position of the point is required.

Example $18.324 \div 0.09 \rightarrow \dfrac{18.324}{0.09} \xrightarrow{\times \text{ top and bottom by 100}} \dfrac{1832.4}{9} \rightarrow \underline{\underline{203.6}}$

● Common fraction arithmetic

Common fraction to decimal fraction

Example $\frac{3}{5} \rightarrow 3 \div 5 \rightarrow 5\overline{)3.0}^{\,0.6} \rightarrow \underline{0.6}$

Decimal fraction to common fraction

Example $0.375 \xrightarrow[\substack{\text{last figure is in the}\\\text{thousandths' column}}]{} \dfrac{375}{1000} \rightarrow \dfrac{\cancel{375}^{\,75\;3}}{\cancel{1000}_{200\;8}} \rightarrow \underline{\underline{\tfrac{3}{8}}}$

Addition and subtraction
It is best to deal with the whole numbers first.

Examples $6\frac{3}{8} + 1\frac{1}{4} \rightarrow 7\frac{3}{8} + \frac{2}{8} \rightarrow \underline{7\tfrac{5}{8}}$

$4\frac{1}{3} - 2\frac{2}{5} \rightarrow 2\frac{5}{15} - \frac{6}{15} \rightarrow 2 - \frac{1}{15} \rightarrow \underline{\underline{1\tfrac{14}{15}}}$

Multiplication of a fraction by an integer

Examples $\dfrac{^2\cancel{8} \times \cancel{9}^3}{_{13}\cancel{12}} \rightarrow \underline{\underline{6}}$

$2\frac{1}{6} \times 3 \rightarrow \dfrac{13 \times \cancel{3}^1}{_2\cancel{6}} \rightarrow \dfrac{13}{2} \rightarrow \underline{\underline{6\tfrac{1}{2}}}$

Fraction multiplied by fraction
Change all mixed numbers to improper (top-heavy) fractions first.

Example $3\frac{3}{4} \times 1\frac{1}{5} \rightarrow \dfrac{^3\cancel{15}}{_2\cancel{4}} \times \dfrac{\cancel{6}^3}{\cancel{5}_1} \rightarrow \dfrac{9}{2} \rightarrow \underline{\underline{4\tfrac{1}{2}}}$

Fraction divided by integer
To divide by n, multiply by its reciprocal $\left(\dfrac{1}{n}\right)$.

Example $\frac{3}{4} \div 4 \rightarrow \frac{3}{4} \times \frac{1}{4} \rightarrow \underline{\underline{\tfrac{3}{16}}}$

Division by a fraction
To divide by a fraction, multiply by its inverse.

Examples $3 \div \frac{1}{2} \rightarrow 3 \times \dfrac{2}{1} \rightarrow \underline{\underline{6}}$

$2\frac{1}{2} \div 1\frac{2}{3} \rightarrow \dfrac{5}{2} \div \dfrac{5}{3} \rightarrow \dfrac{^1\cancel{5}}{2} \times \dfrac{3}{\cancel{5}_1} \rightarrow \dfrac{3}{2} \rightarrow \underline{\underline{1\tfrac{1}{2}}}$

● Divisibility

A number divides exactly by:	2	3	5	6	9	10
if its digit-sum is:	any	3; 6; 9	any	3; 6; 9	9	any
and its last digit is:	even	any	0; 5	even	any	0

A number divides exactly by 4 if its last two digits divide exactly by 4.

A number divides exactly by 8 if its last 3 digits divide exactly by 8.

● Arithmetic with directed numbers

See the notes in Chapter 2 (page 10).

● Approximation

See the notes in Chapter 6 (page 41).

● Prime numbers

Primes have only two different factors; e.g. 19 is prime because its factors are 1 and 19; 9 is not prime because it has three factors, 1, 3 and 9.

To write a number as a product of prime factors:

Example $162 \rightarrow 2 \times 81 \rightarrow 2 \times 3 \times 27 \rightarrow 2 \times 3 \times 3 \times 9 \rightarrow 2 \times 3 \times 3 \times 3 \times 3$

This may be written: 2)162
3) 81
3) 27
3) 9
3) 3
     1

● Highest Common Factor (HCF)

A factor divides exactly into a number. The HCF is the highest factor that divides exactly into a set of numbers. For large numbers a prime factor method is useful.

Examples The HCF of {12, 15, 18} is 3 (this can be done by just thinking about it).

To find the HCF of 168 and 180:

$168 \rightarrow 2 \times 2 \times 2 \times 3 \times 7$

$180 \rightarrow 2 \times 2 \times 3 \times 3 \times 5$

HCF $= 2 \times 2 \times 3 = \underline{\underline{12}}$

● Lowest Common Multiple (LCM)

A multiple is made by multiplying by an integer. The LCM is the lowest number that is a multiple of each member of a given set of numbers.

Examples   The LCM of {6, 8, 12} is 24 (this can be done by just thinking about it).

To find the LCM of 18, 30 and 36:
As 36 is a multiple of 18, we need not think about the 18. All multiples of 30 end in a zero, therefore the answer is a multiple of 36 that ends in a zero and also divides exactly by 30. The answer is 180.

For large numbers a prime factor method is useful.

Example   To find the LCM of {18, 24, 64}.
$18 \rightarrow 2 \times 3 \times 3$
$24 \rightarrow 2 \times 2 \times 2 \times 3$
$64 \rightarrow 2 \times 2 \times 2 \times 2 \times 2 \times 2$
The prime factors of the LCM will consist of 2's and 3's. We need two 3's for 18 and six 2's for 64.
Hence the LCM is $2 \times 2 \times 2 \times 2 \times 2 \times 2 \times 3 \times 3 = 576$.

Test yourself

1   Add: 3000,   8.2,   0.49,   1.099

2   Find the difference between 16 and 9.0974

3   Find the product of 8.76 and 0.97

4   $3.83819 \div 1.9$

5   $3\frac{5}{8} - 1\frac{3}{4}$

6   $1\frac{3}{4} \div 7$

7   $2\frac{2}{3} \div 2\frac{2}{9}$

Questions 8 to 14 provide extra practice on questions like 1 to 7. Your teacher will tell you if you may omit them.

*8   (a) $2.6 + 300 + 0.49$ (b) $9.79 + 9009 + 8.19 + 0.099$

*9   Find the difference between:
(a) 12 and 6.078   (b) 10 and 0.919

*10 Find the product of:
(a) 0.3 and 0.4   (b) 0.1 and 0.07   (c) 10.3 and 0.75   (d) 9.79 and 89
(e) 3000 and 4.01

*11 (a) 57.114 ÷ 1.9 (b) 195.1313 ÷ 1.3 (c) 103.2019 ÷ 1.7 (d) 0.208 84 ÷ 2.3

*12 (a) $4\frac{7}{16} - \frac{3}{8}$ (b) $2\frac{4}{5} - \frac{3}{7}$ (c) $2\frac{5}{9} - 1\frac{5}{6}$ (d) $4\frac{5}{16} - 2\frac{7}{12}$ (e) $3\frac{23}{33} - 1\frac{8}{11}$

*13 (a) $9 \div 2\frac{1}{2}$ (b) $6 \div \frac{1}{4}$ (c) $\frac{2}{3} \div 8$ (d) $1\frac{3}{5} \div 12$ (e) $3\frac{3}{5} \div 9$

*14 (a) $2\frac{1}{2} \div 3\frac{3}{4}$ (b) $3\frac{3}{8} \div 2\frac{1}{4}$ (c) $2\frac{6}{11} \div 2\frac{7}{22}$ (d) $5\frac{1}{7} \div 6\frac{3}{4}$

15 (a) $4 - 12$ (b) 5×-6 (c) -4×-3 (d) $-8 - -5$ (e) $9 - 11$
 (f) -7×-3 (g) $9 - -2$ (h) $-6 + -3$

16 Write a number more than 100 which divides exactly by:
 (a) 2 and 3 but not 9 (b) 6 and 9 (c) 2, 3, 5, 6 and 9 (d) neither 2, 3, 5, 6 or 9.

 Check your answers with a calculator.

17 Find the HCF and LCM of:
 (a) 3, 6 and 15 (b) 24, 32, 48 and 60.

18 Rewrite {7.882, 691.398, 7108.8017} with the elements rounded to:
 (a) 2 significant figures (b) 2 decimal places.

19 In a sale all goods are '12p in the £ off!!' What is the price in the sale of an article normally costing
 (a) £8 (b) £6.50 (c) £10.25?

20 Change the following to the units given in brackets.

 Example 708 mm (cm) *Answer* 70.8 cm

 (a) 35 mg (cg) (b) 7.8 cm (mm) (c) 350 m (km) (d) 3000 cl (litres)
 (e) 5 tonnes (kg) (f) 4.65 km (m) (g) 0.75 cm (mm) (h) 0.19 m (cm)
 (i) 0.5 km (cm) (j) 1 litre (cm^3) (k) 1 m^2 (cm^2) (l) 1 m^3 (cm^3)

21 A house is rated at £380. If the rate declared is 96p in the pound, how much will the householder have to pay?

22 A 12-hour clock loses 75 seconds a day (24 hours). How many days will it take to appear correct again after having been set to the right time?

23 A man's salary is £15 420 a year. He is paid monthly and taxed at 30%, with a tax-free allowance of £2880 a year. He pays insurance of £15 a month plus 2% on gross income. Calculate his nett monthly pay.

24 I buy £150 worth of Italian lire at L2150 to the pound. Three months later, when the exchange rate has changed to L2000 to the pound, I change my lire back to sterling. What profit do I make if I have to pay charges of 1% on the sterling value of both transactions?

25 Light travels at approximately 3×10^5 km/s.

 (a) Find how far a spaceship travelling at a third of the speed of light travels in an hour.

 (b) How long would the spaceship take from Earth to Saturn, a distance of 1.275×10^9 km?

26 A jar of water weighs 6.5 kg when full and 3.5 kg when half full. 1 cm³ of water has a mass of 1 g.

When a quarter full:
(a) what does it weigh   (b) how many litres does it contain?

27 Three carpenters and two apprentices together earn £900 a week. A carpenter's wage to an apprentice's is in the ratio 2 : 1. What does an apprentice earn a week?

28 A swimming-pool is 100 metres long, 60 metres wide and slopes along the length to give a steady increase in depth from 1 metre to 4 metres. How many litres of water will the pool hold? (Answer in standard form.)

29 A rectangular piece of paper is, to the nearest mm, 7.6 cm long and 2.8 cm wide. What is the maximum possible length, width, and area of the paper?

30 A roll of paper on a cylindrical former is 10 cm in diameter. The former is 2 cm in diameter and the paper is 0.1 mm thick.

(a) How many turns of paper are there?

(b) How long is the complete roll of paper? (Take π = 3.1)

31 'DECIFRAC' is a computer program which changes a decimal fraction into a common fraction. The way it does this is hard to understand, but try a 'dry run' using A = $1\frac{3}{4}$ and recording what happens to the values of B, C, D, E, F, G and C(B).

```
  5 REM "DECIFRAC"
 10 DIM C(12)
 20 LET D = 1
 30 LET F = 1
 40 PRINT "Type any decimal number."
 50 INPUT A
 60 PRINT TAB(10);A
 70 FOR B = 1 TO 12
 80 LET C(B) = INT(A + 0.00001)
 90 LET D = B
100 IF ABS(C(B) − A) < 0.00001 THEN LET B = 12
110 LET A = 1/(A − C(B))
120 NEXT B
130 LET E = C(D)
140 FOR B = D−1 TO 1 STEP −1
150 LET G = E
160 LET E = C(B)*E + F
170 LET F = G
180 NEXT B
190 PRINT "ˌas near as I can get it isˌ"
200 PRINT TAB(10); E
210 PRINT TAB(10); "—"
220 PRINT TAB(10); F
```

```
32    1 REM "SEQUENCE"
     10 DIM F(3)
     20 RANDOMIZE (May not be needed)
     30 CLS (Clear screen)
     40 FOR E = 1 TO 3
     50 LET F(E) = INT(RND*3) + 1 (You may need RND(1))
     60 NEXT E
     70 LET B = INT(RND*3) + 1
     80 FOR A = 1 TO 5
     90 PRINT F(1)
    100 IF B = 1 THEN LET F(1) = F(1) + F(2) - F(1)*F(3)
    110 IF B = 2 THEN LET F(1) = F(1)*F(2) - F(3)
    120 IF B = 3 THEN LET F(1) = F(1)*2 - F(2)*3
    130 NEXT A
    140 LET Z = 0
    150 PRINT "?"
    160 INPUT X
    170 IF X = F(1) THEN GOTO 250
    180 IF Z = 1 THEN GOTO 230
    190 PRINT "No. It is between ";F(1) - B - 2;" and "; F(1) + 8 - B
    200 PRINT "Try again."
    210 LET Z = 1
    220 GOTO 150
    230 PRINT "It was "; F(1)
    240 IF Z = 1 THEN GOTO 260
    250 PRINT "Correct. Well done."
    260 LET Z = 0
    270 PRINT "Type any key to continue."
    280 INPUT A$
    290 GOTO 30
```

33 There is a rule for creating an odd (e.g. 3 by 3) magic square. The following program uses this rule. Find out how it works. Investigate other magic squares.

```
 10 REM "MAGICSQ"
 20 PRINT "Type order of square (how many rows it has)."
 30 PRINT "NOTE THAT YOUR NUMBER MUST BE ODD"
 40 INPUT A
 50 IF INT(A/2) = A/2 THEN GOTO 40
 60 DIM B(A,A)
 70 LET T = 0
 80 LET C = 1
 90 LET F = A
100 LET E = (A + 1)/2
110 IF B(E, F) = 0 THEN GOTO 140 (Checks place is not occupied)
120 LET E = E1 (Place next number on same row as previous one)
130 LET F = F1 - 1 (Place next number to left of the previous one)
140 LET B(E, F) = C (Allocate number C to row E, column F)
150 LET E1 = E (Store row of number C)
160 LET F1 = F (Store column of number C)
```

```
170 IF C = A*A THEN GOTO 290 (Print the completed square)
180 LET C = C + 1 (Next highest number)
190 REM Lines 200 to 280 select position of number
200 IF F < > A THEN GOTO 230
210 LET F = 1 (Place in far left column)
220 GOTO 240
230 LET F = F + 1 (Go forward one column)
240 IF E < > A THEN GOTO 270
250 LET E = 1 (Place on top row)
260 GOTO 110 (Position accepted)
270 LET E = E + 1 (Move down a row)
280 GOTO 110 (Position accepted)
290 REM Lines 300 to 420 print the square
300 CLS (Clear screen)
310 FOR E = 1 TO A
320 FOR F = 1 TO A
330 LET L = LEN(STR$(B(E, F)))
340 PRINT TAB (4*F − L); B(E,F);
350 NEXT F
360 PRINT
370 PRINT
375 PRINT (May not be needed)
380 NEXT E
390 FOR E = 1 TO A
400 LET T = T + B(1,E)
410 NEXT E
420 PRINT "Lines total ^"; T
```

12 Matrices: inverse

A Identity and inverse

For a given number system, the **identity** is the element that has no apparent effect under a given operation.

For integers: The identity for addition is 0, e.g. 6 + 0 = 6.
The identity for multiplication is 1, e.g. 6 × 1 = 6.

For 2 by 2 matrices: The identity for addition is $\begin{pmatrix} 0 & 0 \\ 0 & 0 \end{pmatrix}$,

$$\text{e.g.} \quad \begin{pmatrix} 6 & 5 \\ 4 & 3 \end{pmatrix} + \begin{pmatrix} 0 & 0 \\ 0 & 0 \end{pmatrix} = \begin{pmatrix} 6 & 5 \\ 4 & 3 \end{pmatrix}.$$

The identity for multiplication is $\begin{pmatrix} 1 & 0 \\ 0 & 1 \end{pmatrix}$,

$$\text{e.g.} \quad \begin{pmatrix} 1 & 0 \\ 0 & 1 \end{pmatrix} \begin{pmatrix} 6 & 5 \\ 4 & 3 \end{pmatrix} = \begin{pmatrix} 6 & 5 \\ 4 & 3 \end{pmatrix}.$$

For a given number system, the **inverse** of an element 'undoes' the effect of that element for a given operation. The inverse combines with the element to give the identity.

For integers: The inverse of n for addition is $-n$,
e.g. **6 + 3 = 9** and **9 + −3 = 6**.

The inverse of n for multiplication is $\frac{1}{n}$,
e.g. **6 × 3 = 18** and **18 × $\frac{1}{3}$ = 6**.

For 2 by 2 matrices: The inverse of $\begin{pmatrix} a & b \\ c & d \end{pmatrix}$ for addition is $\begin{pmatrix} -a & -b \\ -c & -d \end{pmatrix}$,

$$\text{e.g.} \quad \begin{pmatrix} 6 & 5 \\ 4 & 3 \end{pmatrix} + \begin{pmatrix} -1 & 0 \\ 7 & 9 \end{pmatrix} = \begin{pmatrix} 5 & 5 \\ 11 & 12 \end{pmatrix} \text{ and}$$

$$\begin{pmatrix} 5 & 5 \\ 11 & 12 \end{pmatrix} + \begin{pmatrix} 1 & 0 \\ -7 & -9 \end{pmatrix} = \begin{pmatrix} 6 & 5 \\ 4 & 3 \end{pmatrix}.$$

The inverse of $\begin{pmatrix} a & b \\ c & d \end{pmatrix}$ for multiplication is $\frac{1}{ad - bc} \begin{pmatrix} d & -b \\ -c & a \end{pmatrix}$.

$(ad - bc)$ is called the **determinant** of the matrix $\begin{pmatrix} a & b \\ c & d \end{pmatrix}$.

The example below shows step-by-step how to calculate the multiplicative inverse.

The additive inverse matrix is little used, and the multiplicative inverse is always the one implied when you are asked to find 'the' inverse.

Example To find the inverse of $\begin{pmatrix} 5 & 1 \\ 8 & 2 \end{pmatrix}$.

Step One Write the standard format: $\dfrac{1}{}\begin{pmatrix} & \\ & \end{pmatrix}$

Step Two Calculate the determinant. This is best done by writing it above the original matrix:

$$\overset{10\quad-\quad8\,=\,2}{\begin{pmatrix} 5 & 1 \\ 8 & 2 \end{pmatrix}} \rightarrow \frac{1}{2}\begin{pmatrix} & \\ & \end{pmatrix}$$

Step Three Swap over the elements on the leading diagonal:

$$\begin{pmatrix} 5 & \\ & 2 \end{pmatrix} \rightarrow \frac{1}{2}\begin{pmatrix} 2 & \\ & 5 \end{pmatrix}$$

Step Four Change the signs on the other diagonal, giving the complete inverse:

$$\frac{1}{2}\begin{pmatrix} 2 & -1 \\ -8 & 5 \end{pmatrix}$$

Note: The $\frac{1}{2}$ may be multiplied through to give $\begin{pmatrix} 1 & -\frac{1}{2} \\ -4 & 2\frac{1}{2} \end{pmatrix}$.

The following computer program calculates the inverse of a matrix. Much of the complexity of the program is caused by the difficulty of arranging a tidy display in most BASICs. The actual inverse is worked out in lines 15Ø to 17Ø.

```
  5 REM "INVERMAT"
 1Ø PRINT "Type in a 2 by 2 matrix, using RETURN between each number."
 2Ø INPUT A
 3Ø LET W = LEN(STR$(A))
 4Ø INPUT B
 5Ø LET X = LEN(STR$(B))
 6Ø INPUT C
 7Ø LET Y = LEN(STR$(C))
 8Ø INPUT D
 9Ø LET Z = LEN(STR$(D))
1ØØ PRINT "(";TAB(5 − W); A;TAB(9 − X); B;TAB(11);")"
11Ø PRINT "(";TAB(5 − Y);C;TAB(9 − Z);D;TAB(11);")"
12Ø PRINT
13Ø IF A∗D = B∗C THEN GOTO 19Ø
14Ø PRINT "Multiplicative inverse is ‸"
15Ø LET X$ = STR$(A∗D−B∗C)
16Ø PRINT TAB(LEN(X$)/2);"1";TAB(4);"(";TAB(8−Z);D;
    TAB(12−LEN(STR$(B∗−1)));B∗−1;TAB(14);")"
```

```
170 PRINT; A*D−B*C;TAB(4);"(";TAB(8−LEN(STR$(C*−1)));
    C*−1;TAB(12−W);A;TAB(14);")"
180 GOTO 200
190 PRINT "No inverse. Determinant is zero."
200 PRINT
210 GOTO 10
```

1 Find the inverses of the following matrices.

(a) $\begin{pmatrix} 5 & 2 \\ 7 & 3 \end{pmatrix}$ (b) $\begin{pmatrix} 5 & 3 \\ 3 & 2 \end{pmatrix}$ (c) $\begin{pmatrix} 5 & 7 \\ 2 & 3 \end{pmatrix}$ (d) $\begin{pmatrix} 5 & 4 \\ 2 & 2 \end{pmatrix}$ (e) $\begin{pmatrix} 4 & 2 \\ 1 & 1 \end{pmatrix}$

*2 Find the inverses of the following matrices.

(a) $\begin{pmatrix} 5 & 14 \\ 1 & 3 \end{pmatrix}$ (b) $\begin{pmatrix} 2 & 1 \\ 11 & 6 \end{pmatrix}$ (c) $\begin{pmatrix} 3 & 2 \\ 7 & 5 \end{pmatrix}$ (d) $\begin{pmatrix} 8 & 7 \\ 1 & 1 \end{pmatrix}$

(e) $\begin{pmatrix} 3 & 1 \\ 4 & 2 \end{pmatrix}$ (f) $\begin{pmatrix} 4 & 3 \\ 2 & 2 \end{pmatrix}$ (g) $\begin{pmatrix} 3 & 5 \\ 2 & 4 \end{pmatrix}$ (h) $\begin{pmatrix} 5 & 6 \\ 2 & 3 \end{pmatrix}$

3 Because $\dfrac{1}{0}$ has no real meaning, there is no inverse matrix if the determinant is zero.

Find, where possible, the inverse matrix of:

(a) $\begin{pmatrix} 3 & -4 \\ 2 & -2 \end{pmatrix}$ (b) $\begin{pmatrix} 3 & -4 \\ 3 & -2 \end{pmatrix}$ (c) $\begin{pmatrix} 3 & 9 \\ 2 & 6 \end{pmatrix}$ (d) $\begin{pmatrix} 2 & -5 \\ 1 & 1 \end{pmatrix}$ (e) $\begin{pmatrix} -6 & -3 \\ 3 & 1 \end{pmatrix}$

(f) $\begin{pmatrix} 4 & -2 \\ 7 & -3 \end{pmatrix}$ (g) $\begin{pmatrix} 4 & -6 \\ 2 & -3 \end{pmatrix}$ (h) $\begin{pmatrix} -4 & -2 \\ -5 & -3 \end{pmatrix}$ (i) $\begin{pmatrix} -6 & 5 \\ -1 & -2 \end{pmatrix}$

4 You may need to refer back to Chapter 1 when answering this question.

Plot a simple shape on a grid; transform it by multiplying by a 2 by 2 matrix. Plot the resulting shape. Now transform this shape by multiplying by the inverse of your original transforming matrix. Investigate further.

5 Plot the 'unit square', represented by the matrix $\begin{pmatrix} 0 & 1 & 1 & 0 \\ 0 & 0 & 1 & 1 \end{pmatrix}$. Transform it by multiplying by any 2 by 2 matrix. The area of the unit square is 1 square unit. Calculate the area of your transformed shape. Also calculate the determinant of your transforming matrix. Investigate further.

6 Investigate the commutativity of:
(a) any two 2 by 2 matrices
(b) a 2 by 2 matrix and its multiplicative inverse.

B Simultaneous equations

The Teachers' Manual to Book 3 contains a computer program, SIMUL, which solves simultaneous equations. It uses the following matrix method, which many students find easier than the traditional algebraic approach.

Example To solve $5x + 3y = 11$ and $3x + 2y = 7$ simultaneously.

The equations may be written in a matrix form (rather cunning, this!):

$$\begin{pmatrix} 5 & 3 \\ 3 & 2 \end{pmatrix} \begin{pmatrix} x \\ y \end{pmatrix} = \begin{pmatrix} 11 \\ 7 \end{pmatrix} \quad \text{(Multiply this out to check it.)}$$

$$\therefore \begin{pmatrix} x \\ y \end{pmatrix} = \begin{pmatrix} 11 \\ 7 \end{pmatrix} \div \begin{pmatrix} 5 & 3 \\ 3 & 2 \end{pmatrix}$$

To divide by a matrix we pre-multiply by its inverse (compare this with division by a fraction):

$$\therefore \begin{pmatrix} x \\ y \end{pmatrix} = \frac{1}{1} \begin{pmatrix} 2 & -3 \\ -3 & 5 \end{pmatrix} \begin{pmatrix} 11 \\ 7 \end{pmatrix} = \begin{pmatrix} 1 \\ 2 \end{pmatrix}$$

giving $x = 1$ and $y = 2$.

1 Solve simultaneously:
  (a) $3x + y = 11$ and $2x + y = 9$
  (b) $4x + y = 6$ and $3x + y = 5$

2 **Example** To solve $4x + 5y = 14$ and $2x + 3y = 8$ simultaneously.

$$\begin{pmatrix} 4 & 5 \\ 2 & 3 \end{pmatrix} \begin{pmatrix} x \\ y \end{pmatrix} = \begin{pmatrix} 14 \\ 8 \end{pmatrix} \rightarrow \begin{pmatrix} x \\ y \end{pmatrix} = \frac{1}{2} \begin{pmatrix} 3 & -5 \\ -2 & 4 \end{pmatrix} \begin{pmatrix} 14 \\ 8 \end{pmatrix}$$

$$\rightarrow \begin{pmatrix} x \\ y \end{pmatrix} = \frac{1}{2} \begin{pmatrix} 2 \\ 4 \end{pmatrix}$$

$$\rightarrow \begin{pmatrix} x \\ y \end{pmatrix} = \begin{pmatrix} 1 \\ 2 \end{pmatrix}$$

Answer: $x = 1$ and $y = 2$.

Solve simultaneously:
  (a) $3x + 2y = 13$ and $4x + 3y = 18$ (b) $x + 2y = 8$ and $2x - 3y = 2$
  (c) $2x - 3y = 4$ and $x - y = 3$ (d) $x + 2y = 12$ and $2x - 3y = -11$
  (e) $2x + 3y = -3$ and $4x - y = 8$ (f) $x + y = -6$ and $3x - y = -4$.

3 The following pairs of equations cannot be solved simultaneously. Try to solve them by the matrix method, then draw graphs to show why there are no solutions.
  (a) $2x + y = 7$ and $6x + 3y = 20$
  (b) $4x - 2y = 9$ and $2x - y = 4$

C Using your calculator

Monsieur Eiffel's marvellous tower

(*French Government Tourist Office*)

On January 28, 1887, Gustave Eiffel, a French engineer, started to build his famous Parisian tower. The last rivet was hammered home on March 10, 1889.

The tower starts 14 metres underground. After ten months' work it reached the first deck, 58 metres above ground. A further four months saw the second deck in place, twice as high. The third deck is at 276 metres. Above this are two very small platforms, the highest at 300.65 metres, from which one can see 80 km on a clear day. There used to be 1710 steps from the ground to the highest deck. These were removed in 1983.

During its first year, 1 968 287 visitors paid six million francs to climb the tower, which covered three-quarters of its construction cost.

The tower is very stable and so well built that no rivet has ever had to be replaced. It is little affected by the wind, the strongest gale only moving it 12 cm out of line at the top, but a hot sun can move it 18 cm.

Every seven years, thirty steeplejacks spend eight months using 40 540 litres of paint to paint its 167 225 m^2 of ironwork, adding 45 tonnes to its weight.

Besides being a famous landmark, the tower has been used as a radio and television mast, as well as an exhibition stage for eccentrics who have ridden bicycles down it (!) and climbed its girders. Once a pilot tried to 'thread the arches' (he failed!) and a tailor demonstrated a raincoat parachute (that failed too!).

Use the above information and your calculator to answer the following questions.

1   What fraction of the visible height of the tower is the underground section?

2   How many months did it take from the completion of the second deck to the completion of the tower?

3   What is the distance between the second and third decks?

4   On a clear day how many times further than one's height above the ground can one see from the top?

5   What was the average distance between the steps, in centimetres?

6   Explain why a hot sun causes the tower to lean over.

7   If no painting sessions have been missed, how many litres of paint have been put on the tower so far?

8   What is the weight of a litre of paint in kilograms?

9   How many centilitres of paint do the painters apply on average to each m²?

10 What was the average entry fee paid by each visitor in the first year of the tower's construction?

11 What did the tower cost to construct?

12 Use the formula:
    time in seconds to fall = twice the height in cm divided by 981
    to find how long a stone dropped from the top takes to hit the ground.

13 Use the formula:
    velocity in km/h = $\sqrt{254 \times \text{distance fallen in metres}}$
    to work out the speed of a stone dropped from the top when it hits the ground.

13 Circles: angles (i)

A   Chord and diameter

For Discussion

 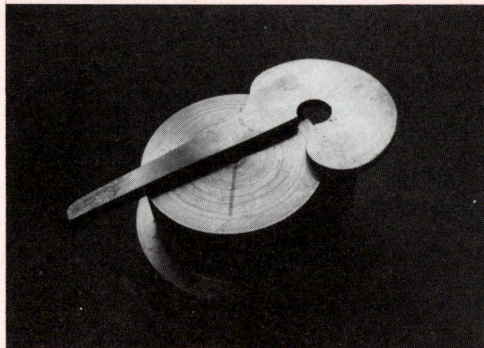

(Robin Cox)

Fig. 13:1

Note: In all the diagrams in this chapter, point O is the centre of the circle.

1 (a) Construct Figure 13:2, using a circle of radius
        20 mm and making AB = 30 mm.

   (b) Join A and B to the centre, O, then mark clearly
        all equal lines and angles.

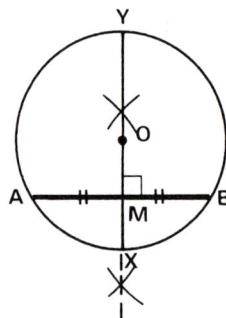

Fig. 13:2

*2 (a) Copy Figure 13:3, using radius about 2 cm.

   (b) If chord AB = 8 cm, write on AM the length it
        would be.

   (c) Copy and complete the following to calculate
        OM if the radius OA = 6 cm.

$$OA^2 = AM^2 + OM^2 \quad \text{(Pythagoras' Theorem)}$$
$$36 = \ldots + OM^2$$
$$\therefore OM^2 = 36 - \ldots = \ldots$$
$$\therefore OM = \sqrt{20} \simeq \ldots \text{ cm}$$

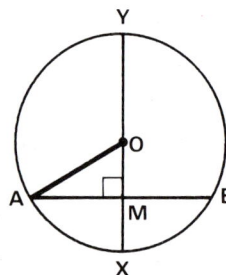

Fig. 13:3

*3   Using Pythagoras' Theorem, as in question 2, and drawing a sketch for each question, calculate for Figure 13:3
     (a) OM if AB = 10 cm and OA = 7 cm
     (b) AM if OM = 3 cm and OA = 8 cm
     (c) OA if AB = 12 cm and OM = 6 cm.

*4   State for Figure 13:3 the length of:
     (a) AB if AM = 7.6 cm
     (b) OA if OX = 4.9 cm
     (c) XY if OA = 10.8 cm.

5   By drawing a suitable small diagram (see Figure 13:3), then using Pythagoras' Theorem (see question 2), calculate:

    (a) the perpendicular distance of an 11 cm chord from the centre of a 7 cm radius circle,

    (b) the length of a chord whose perpendicular distance from the centre of a 5.6 cm radius circle is 3 cm,

    (c) the length of a chord whose perpendicular distance from the centre of an 8.1 cm radius circle is 2.9 cm.

6   Figure 13:4 represents a spherical goldfish-bowl.

    (a) What shape is the surface of the water (AB)?

    (b) Calculate the depth of the water at its deepest point if AB = 12 cm and the bowl has a radius of 10 cm. (Hint: First calculate the perpendicular distance of AB from O.)

    (c) Water is added to the bowl until the water surface has a 12 cm diameter again. How deep is the water now?

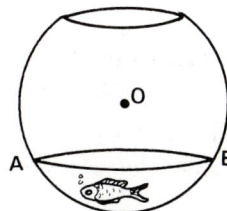

Fig. 13:4

7   Two parallel chords of lengths 6 cm and 8 cm are drawn in a 5 cm radius circle. Calculate the two possible distances between them.

8   Illustrate a method for finding the centre of a circle using two chords and their perpendicular bisectors.

9   Illustrate a method to find the circumcentre of a triangle using the perpendicular bisectors of two chords. Discuss the position of the circumcentre for acute-, right-, and obtuse-angled triangles.

B Angles at centre and circumference

In each of Figures 13:5, 13:6 and 13:7 the arc AB subtends ∠AOB at the centre O and subtends ∠ACB at the circumference.

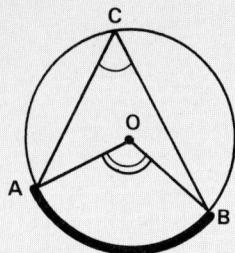

Fig. 13:5

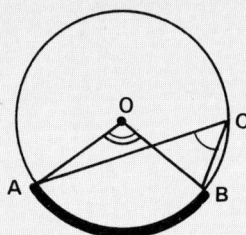

Fig. 13:6

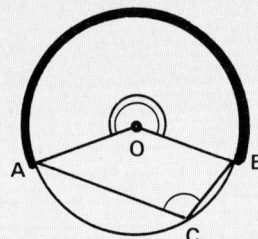

Fig. 13:7

∠AOB = 2∠ACB (Angles at centre and circumference)

Note that in Figure 13:7 it is the major (longer) arc which subtends the angles, and ∠AOB is reflex in this case.

Hints: (a) The pair of angles must start and finish with the same letters, e.g. ∠POQ = 2∠PXR could not be correct.

(b) Both angles must 'open out' the same way. This is especially important when the centre angle is reflex, as in Figure 13:7. This is illustrated in Figure 13:8.

   and   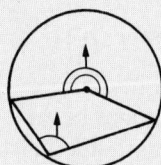   but NOT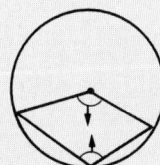

Fig. 13:8

For Discussion

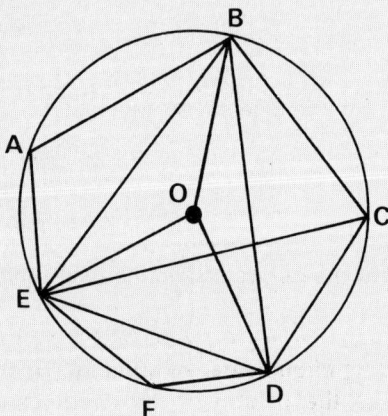

Fig. 13:9

1 Copy Figures 13:10 to 13:13, making them about the same size. Then calculate:
   (a) ∠POQ   (b) ∠POQ reflex   (c) ∠AOC reflex   (d) ∠AOC   (e) ∠XZY
   (f) ∠ZYO   (g) ∠MOK reflex   (h) ∠MNK.

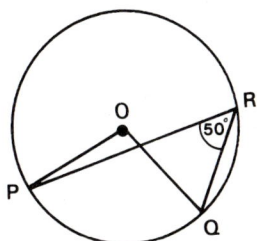

Fig. 13:10

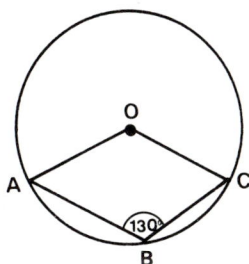

Fig. 13:11

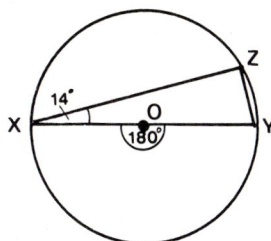

Fig. 13:12

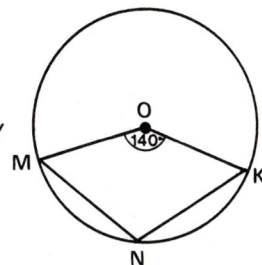

Fig. 13:13

2 Repeat question 1 for Figures 13:14 to 13:17.

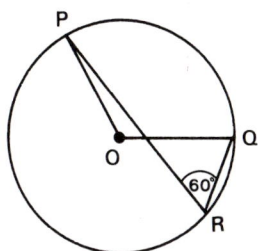

Fig. 13:14

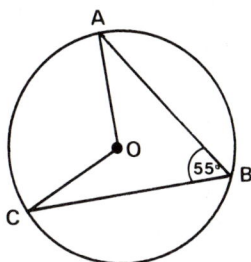

Fig. 13:15

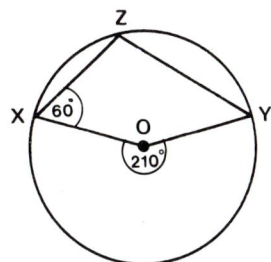

Fig. 13:16

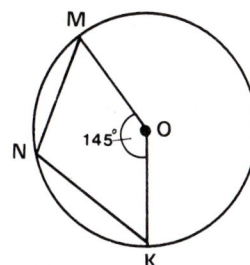

Fig. 13:17

3 For Figures 13:18, 13:19 and 13:20 calculate:
   (a) ∠FDE   (b) ∠HIJ   (c) ∠RNQ   (d) ∠NPQ   (e) ∠PNQ
   (f) ∠NQP   (g) ∠MNR.

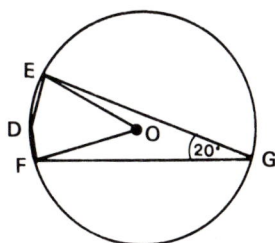

Fig. 13:18

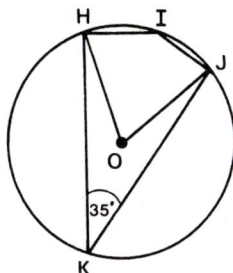

Fig. 13:19

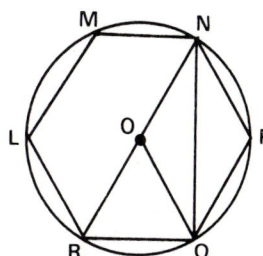

LMNPQR is a regular hexagon

Fig. 13:20

4 On a clock face the points representing 4 and 5 o'clock are joined by straight lines to the point representing 8 o'clock. Calculate the angle between the lines.

5 Calculate the angles of the triangle made by joining the clock face points 3, 8 and 12.

C   Angle in a semicircle

The angle in a semicircle is a right angle

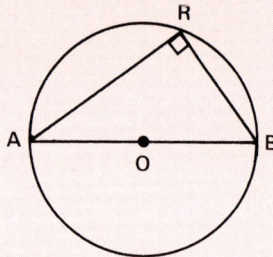

Fig. 13:21

1   In Figure 13:22, which two angles would be 90° if
(a) DF, (b) GE was a diameter?

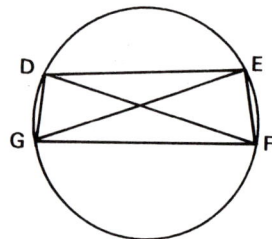

Fig. 13:22

2   Copy Figure 13:23, making it about twice as large.
(a) Name four angles that must be right angles.
(b) If ∠BAF = 47° calculate ∠ABF.
(c) If ∠FAC = 82° calculate ∠FOC.

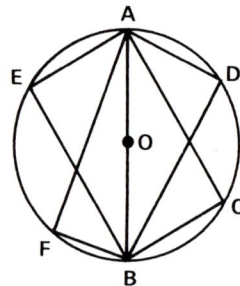

Fig. 13:23

3   (a) Copy Figure 13:24, then calculate the following
angles, giving reasons chosen from (Angle in a
semicircle), (Angles of an isosceles triangle),
(Alternate angles, CB//AD), and (Angle sum of
a triangle).
(i) ∠C   (ii) ∠B   (iii) ∠BAD
(iv) ∠D   (v) ∠AOD

(b) Explain why COD must be a straight line.

(c) Repeat (a) with ∠BAC = 35°.

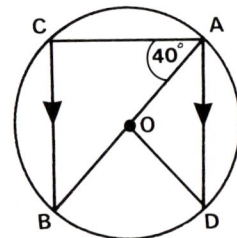

Fig. 13:24

4   Calculate the radius of the circumcircle of the triangle in Figure 13:25.

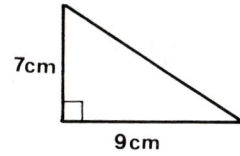

Fig. 13:25

5   In Figure 13:26, find six sets of four concyclic points (i.e. points on the circumference of the same circle); then draw a large copy of the diagram and draw the circles.

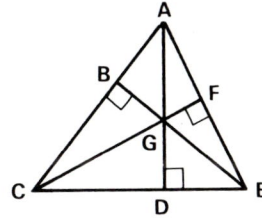

Fig. 13:26

14 Trigonometry: sines and cosines

For Discussion

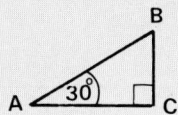
Fig. 14:1

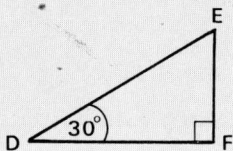
Fig. 14:2

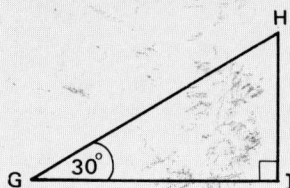
Fig. 14:3

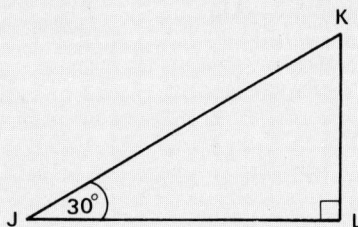
Fig. 14:4

Tangent and Sine

o is the side *o*pposite the angle θ

a is the side *a*djacent to the angle θ

h is the *h*ypotenuse (opposite the right angle)

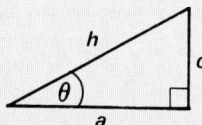

Tangent	$o = a \times \tan\theta$	$\dfrac{o}{a} = \tan\theta$

Sine	$o = h \times \sin\theta$	$\dfrac{o}{h} = \sin\theta$

The following sentence may help you to remember the above facts:

One **a**ncient **t**eacher **o**f **h**istory **s**wore!

1 Example To calculate side o in Figure 14:6.

$o = h \times \sin\theta \rightarrow o = 12 \times \sin 28.7°$

Key: 12 ⊠ 28.7 $\boxed{\text{SIN}}$ $\boxed{=}$

Answer: $o = 5.76$ cm to 3 significant figures.

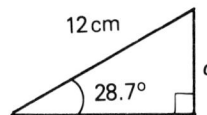
Fig. 14:6

Calculate correct to 3 s.f. the sides marked o in Figure 14:7.

(a)
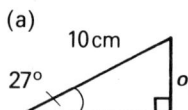

(b)
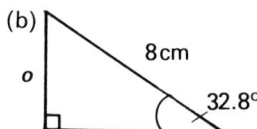

(c)
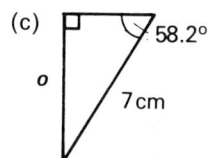

Fig. 14:7

98

2 **Example**   To calculate angle θ in Figure 14:8.

$\dfrac{o}{h} = \sin\theta \rightarrow \dfrac{7}{8} = \sin\theta$

Key: 7 $\boxed{\div}$ 8 $\boxed{=}$ $\boxed{\text{ARCSIN}}$

Answer: $\theta = 61.0°$ to the nearest tenth of 1°.

Fig. 14:8

Note: Your calculator may use INV SIN or SIN^{-1} instead of ARCSIN.

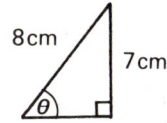

(a) Draw the triangles in Figure 14:9 accurately to the sizes given.

(b) Calculate the angles marked θ, correct to the nearest degree.

(c) Check your answers by measurement with a protractor.

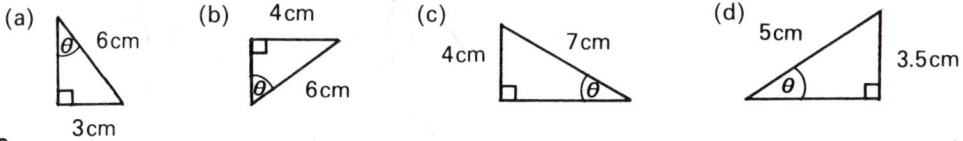

Fig. 14:9

*3** **Example**   To calculate the side x in Figure 14:10.

$8^2 = x^2 + 3^2$ (Pythagoras' Theorem)
$64 = x^2 + 9$
$55 = x^2 \rightarrow x = \sqrt{55} = 7.4$ cm to the nearest 0.1 cm.

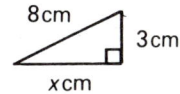

Fig. 14:10

Use Pythagoras' Theorem to calculate the third side of the triangles you drew in question 2. Check your answers by measurement.

*4** Use $o = h \times \sin\theta$ to calculate the side AB in each triangle in Figure 14:11. Remember that θ must be the angle opposite the side that you are trying to find. In some of the triangles, e.g. (a), you are told the size of this angle; in others, e.g. (b), you have to work it out using $\angle A + \angle C = 90°$.

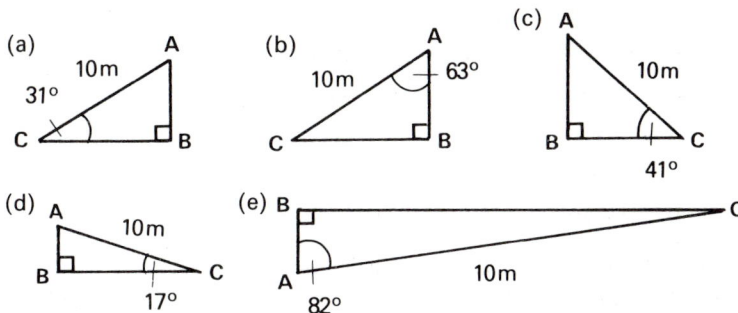

Fig. 14:11

*5** Calculate the sizes of the sides marked x in Figure 14:12.

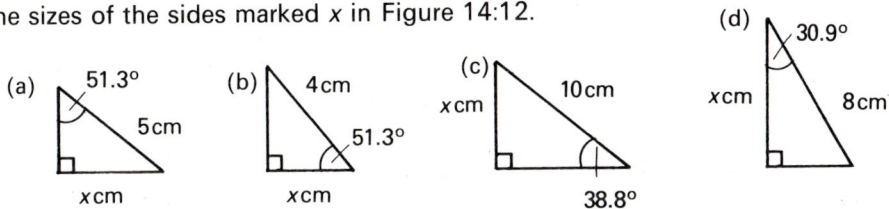

Fig. 14:12

*6 For Figure 14:13
(a) calculate AB using $o = a \times \tan\theta$
(b) calculate AC using Pythagoras' Theorem.

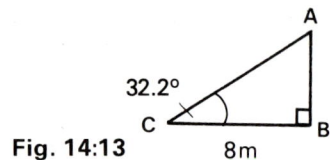
Fig. 14:13

*7 For Figure 14:14
(a) calculate YZ using $o = h \times \sin\theta$
(b) calculate XZ using $o = a \times \tan\theta$.

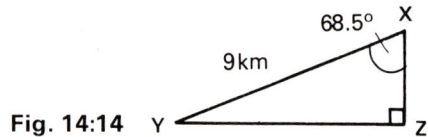
Fig. 14:14

*8 For Figure 14:15
(a) calculate $\angle R$ using the tangent ratio
(b) calculate PR using Pythagoras' Theorem.

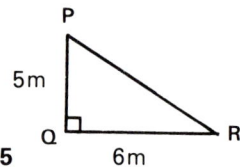
Fig. 14:15

9 For Figure 14:16
(a) calculate $\angle F$ using the sine ratio
(b) use your answer to part (a) to calculate $\angle E$
(c) calculate DF using Pythagoras' Theorem
(d) calculate DF using $\angle E$ and the tangent ratio
(e) calculate DF using $\angle E$ and the sine ratio.

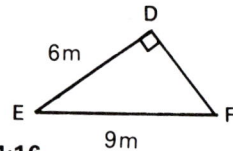
Fig. 14:16

10 The third ratio used in a right-angled triangle is the cosine.

The cosine of an angle is the ratio: adjacent side over hypotenuse.

Learn the complete set of facts set out below.

$o = a \times \tan\theta \qquad o = h \times \sin\theta \qquad a = h \times \cos\theta$

$\dfrac{o}{a} = \tan\theta \qquad\qquad \dfrac{o}{h} = \sin\theta \qquad\qquad \dfrac{a}{h} = \cos\theta$

One ancient teacher of history swore at his class!

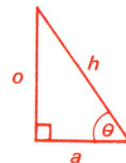

11 (a) For Figure 14:18 find the length of side a correct to 3 s.f.
(b) For Figure 14:19 calculate θ in degrees and minutes, correct to the nearest minute.
(See question 5 of Exercise 8A.)

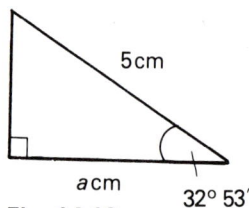
Fig. 14:18

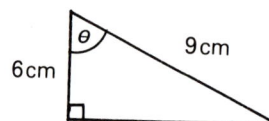
Fig. 14:19

You have probably realised that cosines are not essential to solve a right-angled triangle. Sines could always be used instead, by finding the other angle.

For questions 12 to 14 use sines, cosines, tangents, angle sum of a triangle and symmetry.

12 For Figure 14:20 find:
    (a) ∠BDM   (b) MC   (c) MB
    (d) MD   (e) ∠MCD   (f) ∠MBD
    (g) ∠MAC   (h) ∠MAB   (i) ∠MCA
    (j) ∠MBA.

13 For Figure 14:21 find:
    (a) ∠HEG   (b) ∠FEH   (c) ∠F
    (d) EH to 3 s.f.   (e) HG
    (f) FH (using the answers to parts (b) and (d))
    (g) EF to 3 s.f.

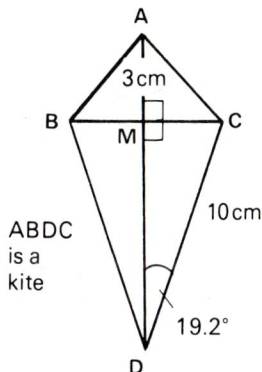

ABDC is a kite

Fig. 14:20

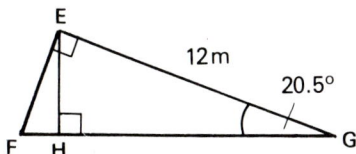

Fig. 14:21

14 For Figure 14:22 calculate all the lengths, and all the acute angles.

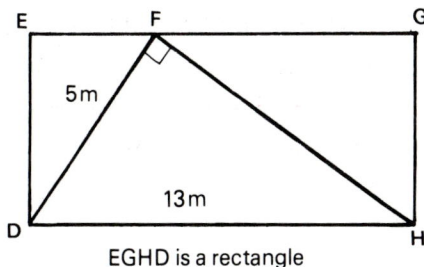

EGHD is a rectangle

Fig. 14:22

15 See Figure 14:23.

    (a) Copy and complete:

        (i) $\theta + \alpha = \ldots°$ (ii) $\sin \theta = \dfrac{AB}{}$ (iii) $\cos \alpha = \dfrac{}{}$

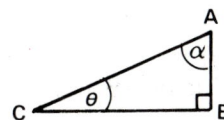

Fig. 14:23

    (b) Using the results of part (a) state the value of θ if:
        (i) $\sin 60° = \cos \theta$ (ii) $\sin \theta = \cos 50°$
        (iii) $\sin 73° \, 18' = \cos \theta$ (iv) $\sin \theta = \cos 80° \, 5'$.

16 For Figure 14:24
    (a) give 3 methods to find x without finding z first
    (b) give 6 methods to find z once you know x.

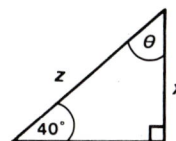

Fig. 14:24

17 Computers and trigonometry

Computers do not use degrees as their angle measure, they use radians. A radian is the angle subtended at the centre of a circle by an arc of the same length as the radius. As $2\pi r$ is the length of the circumference, this means that π **radians** = **180°**. You should remember this fact. The sign for radian is a raised c.

$$1^c = \frac{180°}{\pi} \quad \text{and} \quad 1° = \frac{\pi^c}{180}$$

This will find the tangent of 37°:

10 PRINT TAN (37*PI/180)

You may need to define PI first by

LET PI = 4*ATN(1)

ATN is Basic for arctan. Tan 45° = 1, so ATN(1) = 45° in radians, and 4 * ATN(1) = 180° in radians = PI.

Some computers only have arctan (not arcsin or arccos).
To find arcsin e use:

$$\arcsin e = \arctan\left(\frac{e}{\sqrt{1 - e^2}}\right)$$

To find arccos e use:

$$\arccos e = \arctan\left(\frac{\sqrt{1 - e^2}}{e}\right)$$

These two formulae are obtained from the identities:

$$\tan \theta = \frac{\sin \theta}{\cos \theta} \quad \text{and} \quad \sin^2\theta + \cos^2\theta = 1$$

Try to work them out for yourself.

This will find the angle whose sine is 0.5:

10 PRINT ATN (0.5/SQR(1 − 0.5↑2))*180/PI

15 Expansion of quadratic brackets

A $(x + a)(x + b)$

Examples $(x + 2)(x + 3) \rightarrow x^2 + 3x + 2x + 6 \rightarrow x^2 + 5x + 6$

$(m - 3)(m + 2) \rightarrow m^2 + 2m - 3m - 6 \rightarrow m^2 - m - 6$

Note: The middle step should be done mentally when you have had some practice.

Expand:

1 (a) $(a + 3)(a + 4)$ (b) $(a + 5)(a + 2)$ (c) $(a + 1)(a + 2)$

2 (a) $(p + 5)(p - 3)$ (b) $(m + 6)(m - 2)$ (c) $(p - 5)(p + 2)$
(d) $(r - 7)(r + 4)$ (e) $(p - 2)(p - 4)$ (f) $(a - 7)(a - 4)$

3 (a) $(m + 3)^2$ [That is, $(m + 3)(m + 3)$]   (b) $(a + 4)^2$ (c) $(x - 2)^2$
(d) $(x - 4)^2$

*4 (a) $(b - 4)(b + 5)$ (b) $(a + 2)(a + 8)$ (c) $(a - 5)(a - 3)$ (d) $(a - 8)(a + 4)$

*5 (a) $(r - 7)(r + 4)$ (b) $(p + 6)^2$ (c) $(b - 3)^2$ (d) $(m - 7)^2$

6 **Example** $(4 + x)(3 - x) \rightarrow 12 - 4x + 3x - x^2 \rightarrow 12 - x - x^2$

Note: Do not attempt to change the order of the given terms.

Write down the products of:
(a) $(5 + x)(3 + x)$ (b) $(4 - x)(2 + x)$ (c) $(5 + x)(3 - x)$ (d) $(2 + x)(7 - x)$
(e) $(3 - x)(2 - x)$ (f) $(4 - x)^2$

7 Expand:
(a) $(x + a)(x + b)$ (b) $(x - a)(x + b)$ (c) $(x + a)(x - b)$ (d) $(x - a)(x - b)$

---

8 Use what you have learnt in this exercise to express the unshaded area in terms of x for each diagram in Figure 15:1.

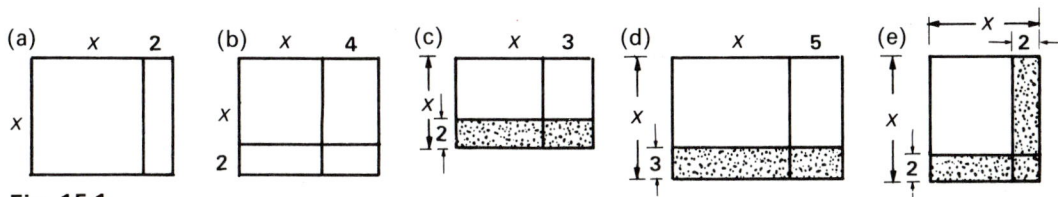

Fig. 15:1

9 Expand:
(a) $(x + 3)(x + a + 1)$ (b) $(x + 2)(x + a - 2)$ (c) $(x - 1)(x + a + 3)$

B $(ax + b)(cx + d)$

Example $(2x + 3)(4x - 5) \rightarrow 8x^2 - 10x + 12x - 15 \rightarrow 8x^2 + 2x - 15$

Expand:

1 (a) $(3x + 5)(x + 2)$ (b) $(2x + 2)(3x + 8)$ (c) $(2x - 4)(3x + 8)$
(d) $(3x - 5)(2x - 3)$

2 (a) $(3x - 3)^2$ (b) $(a - 1)^2$ (c) $(8p - 3)^2$ (d) $(3q + 6)(3q - 6)$

*3 (a) $(2x + 4)(3x + 5)$ (b) $(3x - 2)^2$ (c) $(2x - 4)(3x + 2)$ (d) $(4a - 5)(3a - 2)$

*4 (a) $(2x + 3)^2$ (b) $(x - 4)(3x - 2)$ (c) $(4x - 5)(x + 7)$ (d) $(4x - 5)^2$

5 (a) $(3m + 4n)(2m - 3n)$ (b) $(5a + 2b)(3a - 8b)$ (c) $(4a - 7b)(3a + 2b)$
(d) $(3a - 2b)^2$

---

6 **Example** $102^2 \rightarrow (100 + 2)(100 + 2)$
$\rightarrow 10\,000 + 200 + 200 + 4 \rightarrow 10\,404$

Use the above method to work out:
(a) 105^2 (b) 81^2 (c) 84^2 (d) 109^2 (e) 78^2.

7 Figure 15:2 illustrates $(x + 3)(x + 1) \rightarrow x^2 + 4x + 3$.

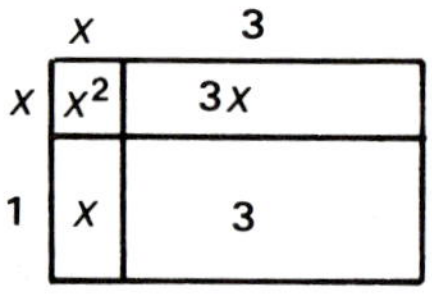

Fig. 15:2

Draw similar diagrams to illustrate:
(a) $(x + 2)(x + 3)$ (b) $(x - 1)(x + 1)$ (c) $(x - 2)(x - 3)$.

16 Graphs: parabolas, $y = x^2 + c$, $y = ax^2$

● You need to know . . .

● Co-ordinates

See the notes on page 1.

● Linear (straight-line) graphs

All straight lines can be expressed in equation form as $y = mx + c$, though the three terms may be moved around, e.g. $y = 2x$, $y = 3$, $y = 2x + 3$, $x + y = 2$, $2y + 3x = 4$ and $x = 2 - 7y$ are all equations of straight-line graphs.

Linear graphs are drawn by one of the following methods.

(a) **Plotting method for $y + 2x = 3$**
Choose three values for x (including zero) and find y for each, e.g.:

If $x = 0$ then $y + 0 = 3 \rightarrow y = 3$. Plot (0, 3).
If $x = 2$ then $y + 4 = 3 \rightarrow y = -1$. Plot (2, −1).
If $x = -1$ then $y - 2 = 3 \rightarrow y = 5$. Plot (−1, 5).

See Figure 16:1.

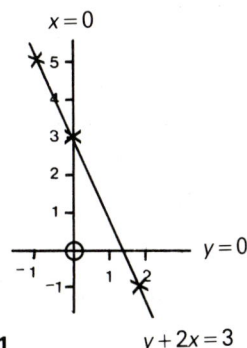

Fig. 16:1 $y + 2x = 3$

(b) **Slope/crossing ($y = mx + c$) method for $y + 2x = 3$**
When the equation is expressed in the form $y = mx + c$ then c gives the crossing point on the y-axis (0, c) and m gives the slope (see Figures 16:2 and 16:3).

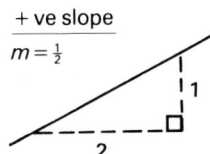

Fig. 16:2

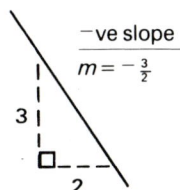

Fig. 16:3

To use this method for $y + 2x = 3$ we first rearrange the terms to change it to $y = -2x + 3$. We now know that $m = -2$ and $c = 3$. Start from 3 on the y-axis, then go down and across to give a slope of -2. See Figure 16:4.

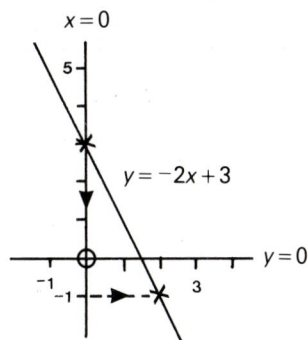

Fig. 16:4

See also the use of linear graphs to solve simultaneous equations, page 12.

● Graph regions

Figure 16:5 shows the line $y = -x + 2$.

For every point on this line the y-co-ordinate is equal to $-x + 2$, where x is the x-co-ordinate of the point.

Above the line, y is more than $-x + 2$.

Below the line, y is less than $-x + 2$.

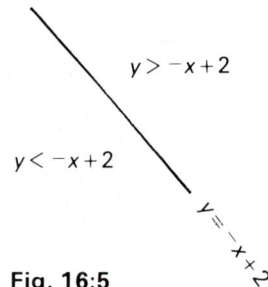

$y > -x + 2$

$y < -x + 2$

$y = -x + 2$

Fig. 16:5

Figure 16:6 shows the region

$\{(x, y): \quad x < 3 ; \quad -x + 2 < y < x\}.$

to the left   above   below
of $x = 3$ $y = -x + 2$ $y = x$

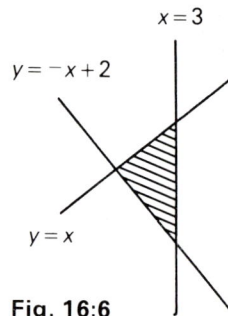

$x = 3$

$y = -x + 2$

$y = x$

Fig. 16:6

Test yourself

1 Draw and label x- and y-axes, each from -5 to 5. Plot the following points.
A, (5, 5)   B, (5, -4)   C, (0, -4)   D, (5, 0)   E, (-2, -4),   F, (3, -4)

2 On your graph for question 1 draw the two straight lines through points A, D, B and through E, C, F, B. Label these lines with their equations.

State the co-ordinates of the point which is equidistant from the origin and points D, B and C.

3 On one set of axes from -5 to 5:
(a) plot the graph whose equation is $y = x - 5$
(b) plot the graph whose equation is $y = -2x$
(c) plot the graph whose equation is $x + y = 3$.

4 On axes from 0 to 3 join the points (0, 1) and (3, 3). State the equation of the line that you have drawn.

5 On axes from -3 to 3 shade the region defined by
$\{(x, y): x < 2; y > -1; -x < y < x\}$

6 Investigate the plotting of points by a computer. Most computers use a simple PLOT X, Y command, although they usually only use the positive quadrant. You have to set up special instructions if you wish to move the origin away from the bottom left-hand corner of the screen. Computers differ too widely in their use of PLOT to give examples here.

Parabolas

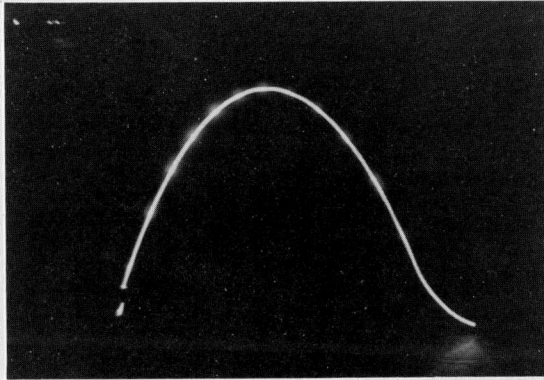

Fig. 16:7 The path of a projectile, like this torch thrown in a dark room, is a parabola

(John R. Bradshaw)

Equations like $y = x^2$, $y = x^2 + 3$ and $y = 2x^2$ give **parabolas** when drawn as graphs.

Sometimes the equation is given in function notation, e.g. $y = f(x)$ where $f(x) : x \rightarrow x^2$. This has the same meaning as $y = x^2$.

To draw the graphs, plot values of y for a series of values of x. (You are usually told which values of x to use.) It is best to work out the values in a table.

Example   To draw the parabola $y = 2x^2 - 3$.

Values for x:

Working out:

Values for y

x	-2	-1	0	1	2
$2x^2$	8	2	0	2	8
-3	-3	-3	-3	-3	-3
y	5	-1	-3	-1	5

Add these two rows to find y.

The points $(-2, 5)$, $(-1, -1)$, etc. are now plotted and joined with a smooth continuous curve. See Figure 16:8.

Graph of $y = 2x^2 - 3$

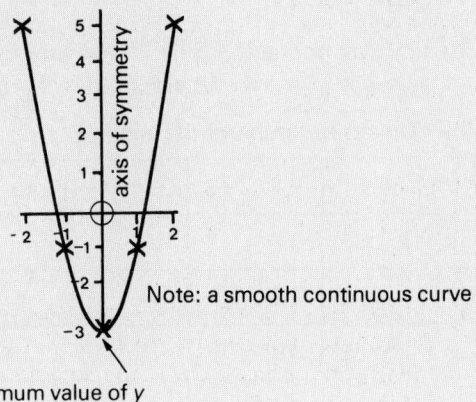

Fig. 16:8   minimum value of y

Note: a smooth continuous curve

1 (a) Copy and complete the table for $y = x^2$:

x	-4	-3	-2	-1	0	1	2	3	4
y	16					1			

(b) On 2 mm graph paper draw x- and y-axes. Use different scales on the two axes as follows: x from -4 to 4, 1 cm to 1 unit; y from 0 to 20, 1 cm to 5 units.

(c) Plot the co-ordinates from the table: $(-4, 16)$, $(-3, 9)$, etc.

(d) Join the points with a smooth curve.
Hints: Work from inside the curve, turning the paper as you go. Use your wrist as a pivot. Watch that you make the turning point curved, not pointed.

(e) Label your graph with its equation and *trace it* for use later in the exercise. (It is very important that you always remember to label your graphs with their equations.)

2 Copy and complete the following table, using the one that you drew up for question 1 to help you.

x	-4	-3	-2	-1	0	1	2	3	4
$y = x^2 + 4$	20	13			4	5			
$y = x^2 - 4$	12	5			-4	-3			
$y = x^2 + 9$	25								
$y = x^2 - 9$	7								

3 On 2 mm graph paper draw axes: x from -4 to 4, 1 cm to 1 unit; y from -10 to 25, 1 cm to 5 units. Plot on this grid the graph of $y = x^2$ together with all four graphs given in question 2, e.g. for $y = x^2 + 4$ plot $(-4, 20)$, $(-3, 13)$, etc. and for $y = x^2 - 4$ plot $(-4, 12)$, $(-3, 5)$, etc.

*4 (a) Check with your tracing of $y = x^2$ that all the graphs plotted in question 3 have the same shape.

(b) You will notice that $y = x^2 + 4$ crosses the y-axis at $y = 4$. Similarly $y = x^2 - 4$ crosses the y-axis at $y = -4$. Where should the graph of $y = x^2 + 3$ cross the y-axis?

(c) Copy and complete the table:

x	-4	-3	-2	-1	0	1	2	3	4
$y = x^2 + 3$	19								
$y = x^2 - 3$	13								
$y = x^2 + 2$	18								
$y = x^2 - 2$	14								

(d) Draw one pair of axes: x from -4 to 4, 1 cm to 1 unit; y from -5 to 20, 1 cm to 5 units. Plot on your grid the graphs of the equations in your table.

(e) Check that each graph crosses the y-axis at the expected point.

*5 Copy and complete the table:

x	-4	-3	-2	-1	0	1	2	3	4
$y = x^2$	16								
$y = -x^2$	-16								
$y = 2x^2$	32								
$y = -2x^2$	-32								
$y = 3x^2$		27							
$y = -3x^2$		-27							

*6 Draw one pair of axes: x from -4 to 4, 1 cm to 1 unit; y from -35 to 35, 1 cm to 5 units. Plot on your grid the graphs of the equations in the table for question 5.

*7 Draw up a table for $y = \frac{1}{2}x^2$, taking x from -4 to 4, then draw the graph of $y = \frac{1}{2}x^2$, choosing sensible scales.

8 (a) By considering the graphs drawn in question 3, state the co-ordinates of the crossing point on the y-axis for:
(i) $y = x^2 + 2$ (ii) $y = x^2 - 5$ (iii) $y = x^2 + a$.

(b) State the equation of the line of symmetry of the family of graphs $y = x^2 + a$.

9 Without using graph paper, and plotting no more than three points, sketch the graph of:
(a) $y = x^2 + 1$ (b) $y = x^2 - 3$.

Show clearly where your graphs cross the y-axis.

10 On one pair of axes: x from -4 to 4, 1 cm to 1 unit; y from 0 to 35, 1 cm to 5 units, draw accurately the graphs of $y = x^2$, $y = 2x^2$, and $y = 3x^2$. (Ignore any values of y outside the given axes' ranges.)

11 On one pair of axes. x from -4 to 4, 1 cm to 1 unit; y from -35 to 0, 1 cm to 5 units, draw accurately the graphs of $y = -x^2$, $y = -2x^2$, and $y = -3x^2$.

12 In Figure 16:9, A, B, C and D are the graphs of the following equations. Which graph has which equation?

$$y = -\tfrac{1}{4}x^2$$
$$y = -\tfrac{1}{2}x^2$$
$$y = x^2$$
$$y = 4x^2$$

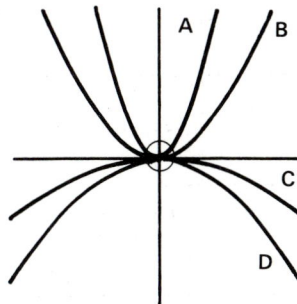

Fig. 16:9

13 Explain how a graph of $y = x^2$ could be used to find an approximate value for $\sqrt{7}$.

14 Draw, for values of x from -3 to 3 at 0.5 intervals, the graph of:

(a) $y = x^3$ (b) $y = -x^3$ (c) $y = \dfrac{1}{x}$ (d) $y = \dfrac{1}{x^2}$ (e) $y = x^3 + 3x^2 - x + 1$.

17 Factorisation: common factors; difference of two squares

A Common factors

If all the terms in an expression have a common factor then the common factor may be 'taken out' and written in front of a bracket.

Examples $3 + 6a \rightarrow 3(1 + 2a)$

$2a - ab \rightarrow a(2 - b)$

$3ax^2 + ax \rightarrow ax(3x + 1)$

Make sure that you take out the *highest* common factor.

1 Copy and complete:
(a) $7 - 14a \rightarrow 7(\quad - \quad)$ (b) $3xy + 4x \rightarrow x(\quad + \quad)$ (c) $16x - 12 \rightarrow 4(\quad - \quad)$
(d) $2x + 4xy \rightarrow 2x(\quad + \quad)$ (e) $3a - 2ab \rightarrow a(\quad - \quad)$ (f) $6ab + 9b \rightarrow 3b(\quad + \quad)$.

*2 Copy and complete:
(a) $mx + my \rightarrow m(\quad)$ (b) $ab + 2b \rightarrow b(\quad)$ (c) $4a - 16 \rightarrow 4(\quad)$
(d) $9a + 3b + 12c \rightarrow 3(\quad)$ (e) $15ab - 10bc + 20bd \rightarrow 5b(\quad)$.

*3 Copy and complete:
(a) $3cd + 6c \rightarrow \quad (d + 2)$ (b) $6ab + 10b \rightarrow \quad (3a + 5)$
(c) $5b - ab \rightarrow \quad (5 - a)$ (d) $2a + ab \rightarrow \quad (2 + b)$
(e) $4a + 6ab \rightarrow \quad (2 + 3b)$ (f) $6a - 9ab \rightarrow \quad (2 - 3b)$.

*4 Factorise:
(a) $6a - 6b$ (b) $7ab - 28$ (c) $4x - 8$ (d) $12 - 8a$ (e) $8b - 4c$ (f) $4a + 8b$
(g) $12 - 3a$ (h) $3bd + 12bc$.

Factorise the expressions in questions 5 to 13 where possible. Where an expression will not factorise, write 'No factors'.

5 (a) $3x - 6xy$ (b) $am - an$ (c) $16abc - 12ab$ (d) $3abx - 6aby$

6 (a) $2x^2 - x$ (b) $3a^2 + 15a$ (c) $2x^2y - 2xy^2$ (d) $12m^2n - 18mn^3$

7 (a) $ab - ca$ (b) $abc - 3bd$ (c) $12y^2 - 8y$ (d) $3ac - 7bd$

8 (a) $4mnp + 6mp$ (b) $6mn^2 - 9m^2n$ (c) $5a^2b - 7c^2d$ (d) $2\pi r^2 + 2\pi rh$

9 (a) $2a^2 + 3b^2$ (b) $24x^2 - 16x$ (c) $21h^2 - 14hn$ (d) $a^2b - ab^3$

10 (a) $4a - 16a^2b$ (b) $3ab - 3a^3b^2$ (c) $amxy - bmdx$

11 (a) $4m - 12n - 6b + 8d$ (b) $6a + 9b - 12c - 3$

12 Example   To factorise $3x - 3y + ax - ay$.

$3x - 3y + ax - ay \rightarrow 3(x - y) + a(x - y)$
Note that $(x - y)$ is itself now a common factor and may be 'taken out':
$3(x - y) + a(x - y) \rightarrow (x - y)(3 + a)$

Multiply $(x - y)(3 + a)$ to check the answer.

(a) $3x + 3y + ax + ay$ (b) $am + 5m + 3a + 15$ (c) $ab + b + 4a + 4$
(d) $5b - 10bc - 2d + 4cd$ (e) $cx + 2cy - dx - 2dy$ (f) $ap + 2bd + ad + 2bp$

13 Example $\dfrac{6x + 6}{2xy + 2y} \rightarrow \dfrac{3(2x + 2)}{y(2x + 2)} \rightarrow \dfrac{3}{y}$

(a) $\dfrac{2x + 6}{xy + 3y}$ (b) $\dfrac{2am + 2ac}{6bm + 6bc}$ (c) $\dfrac{4a^2m + 4ap}{5abm + 5bp}$ (d) $\dfrac{6a^2p + 18ab}{9abp + 27b^2}$

B   Difference of two squares

If two terms in an expression are both squares and are connected by a minus, then they may be split into 'sum times difference'.

Examples $a^2 - 16 \rightarrow (a + 4)(a - 4)$

$4x^2 - 25 \rightarrow (2x + 5)(2x - 5)$

This can be useful in arithmetic.

Examples $54^2 - 46^2 \rightarrow (54 + 46)(54 - 46) \rightarrow 100 \times 8 = 800$

Area of an annulus (e.g. washer) is $\pi R^2 - \pi r^2$.
$\pi R^2 - \pi r^2 \rightarrow \pi(R^2 - r^2) \rightarrow \pi(R + r)(R - r)$

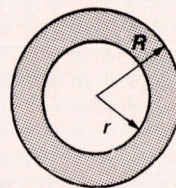

Fig. 17:1

Use this 'difference of two squares' method to factorise the expressions in questions 1 to 6.

1   (a) $a^2 - 9$ (b) $25 - b^2$ (c) $1 - c^2$ (d) $81 - e^2$ (e) $1 - 25f^2$
(f) $4c^2 - 49$

2   (a) $a^2 - 64$ (b) $9a^2 - 25$ (c) $16 - b^2$ (d) $e^2f^2 - 1$ (e) $a^2 - 64b^2$
(f) $16 - 81b^2$

3   Evaluate:
(a) $59^2 - 58^2$ (b) $102^2 - 98^2$ (c) $65^2 - 35^2$ (d) $17^2 - 15^2$ (e) $375^2 - 25^2$
(f) $0.78^2 - 0.22^2$ (g) $367^2 - 133^2$ (h) $1.47^2 - 0.53^2$

17

4 Taking $\pi = 3\frac{1}{7}$ find the area of each annulus in Figure 17:2.

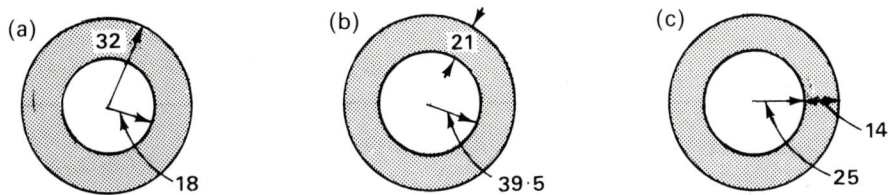

(a) 32 18

(b) 21 39·5

(c) 14 25

Fig. 17:2

5 **Example** $2a^2 - 18 \rightarrow 2(a^2 - 9) \rightarrow 2(a + 3)(a - 3)$

(a) $2a^2 - 32$ (b) $8 - 2x^2$ (c) $27 - 3b^2$ (d) $75 - 3b^2$

6 (a) $a^3 - 9a$ (b) $18a - 2ab^2$ (c) $2\pi r^2 + 2\pi rh$ (d) $25x^2 - 36y^2$
(e) $36x^2 - (x + 1)^2$ (f) $(x + 6)^2 - 36x^2$

18 Circles: angles (ii)

A Angles in the same segment

A **chord** cuts a circle into two **segments**. In Figure 18:1 angles AXB and AYB are **angles in the same segment** (the segment cut off by chord AB). These angles are equal: $\angle AXB = \angle AYB$.

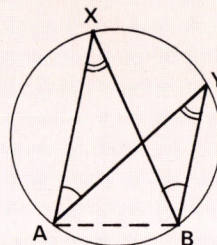

Fig. 18:1

If a chord was drawn from X to Y then angles XAY and XBY would also be 'equal angles in the same segment', though of course it is a different segment to the first one.

Angles in the same segment must start and finish with the same letter, and the middle letter must be a point on the circumference.

For Discussion

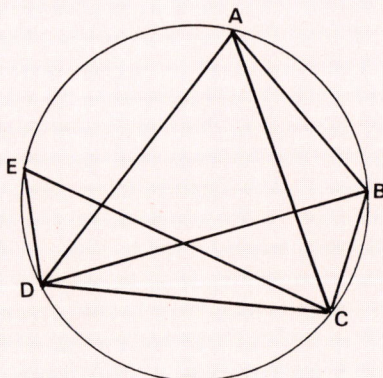

Fig. 18:2

1 In Figure 18:3, state which angle is an equal angle in the same segment as:
(a) angle *x* (b) angle *z*.

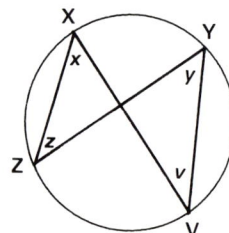

Fig. 18:3

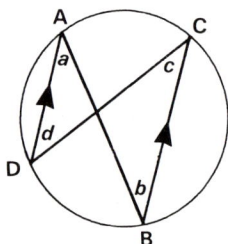

Fig. 18:4

2 In Figure 18:4, AD is parallel to CB. State why the following facts are true:
(a) $a = c$ (b) $a = b$ (c) $b = c$ (d) $b = d$
(e) $c = d$ (f) $a = d$.

*3 Copy Figure 18:5, where XY is parallel to UT.

  (a) State the sizes of the seven angles *a* to *g*.

  (b) Name two pairs of equal alternate angles (use capital letters).

  (c) Name two pairs of equal vertically opposite angles (use small letters).

  (d) Name four pairs of adjacent angles on a straight line (use small letters).

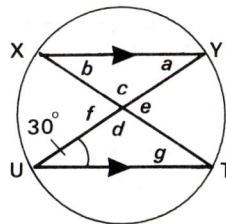

Fig. 18:5

4 In Figure 18:6 name an angle equal to each of the following angles, giving the reason.
  (a) *a*   (b) *b*   (c) *c*   (d) *e*   (e) *i*   (f) *j*

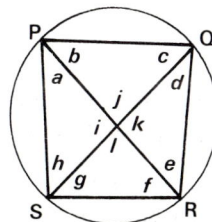

Fig. 18:6

5 Copy Figure 18:7. State with reasons the size of:
  (a) ∠ABC   (b) ∠A   (c) ∠D.

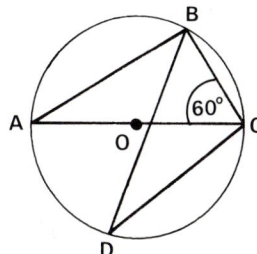

Fig. 18:7

6 In Figure 18:8, XD = XC. State the reason why:
  (a) *c* = *d*
  (b) *d* = *a*
  (c) *a* = *c* (not 'equal alternate angles', as the lines are not known to be parallel)
  (d) AB must be parallel to DC.

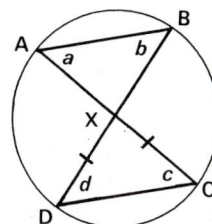

Fig. 18:8

7 In Figure 18:9:
  (a) If ∠ABC = 90° what is AC?
  (b) If X is the centre of the circle then name two angles half the size of ∠BXC.

8 Copy Figure 18:9. If ∠ABD = 50° and ∠ADC = 80° prove that triangle ADC is isosceles.

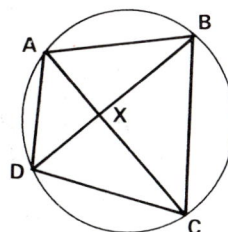

Fig. 18:9

9 In Figure 18:10 state:
(a) the size of ∠ACT
(b) the size of ∠ABC
(c) c + e in degrees
(d) c + a in degrees
(e) the reasons why a = e
(f) the reasons why a = d
(g) the reasons why d = e.

Fig. 18:10

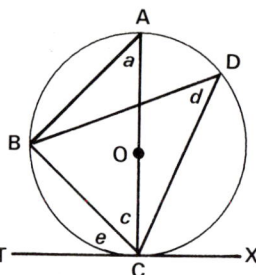

10 ABCD is a cyclic quadrilateral (its corners lie on a circle). Diagonals AC and BD meet at X, and AX = AB. Prove that DX = DC.

B Cyclic quadrilaterals

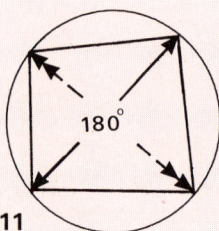

Fig. 18:11

A cyclic quadrilateral has its four corners on the circumference of a circle.

The opposite angles of a cyclic quadrilateral add up to 180° (they are 'supplementary').

1 In Figure 18:12 state the sizes of angles *a* and *b*.

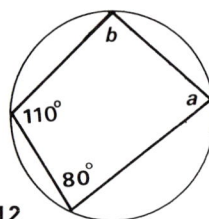

Fig. 18:12

2 In Figure 18:13 AB is parallel to DC. Angles *b* and *c* are interior angles between parallel lines. They add up to 180°. State with the reason the size of:
(a) angle *d*   (b) angle *b*.

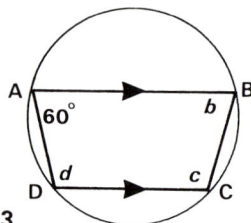

Fig. 18:13

3 Referring to Figure 18:14, copy and complete:
a = 2c (Angles at centre and circumference)
b = . . . (Angles at centre and circumference)
a + b = . . .° (Angles round a point)
∴ 2c + 2d = 360° (a = 2c and b = 2d)
∴ c + d = . . .° (dividing by 2)

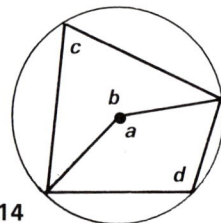

Fig. 18:14

*4 In Figure 18:15
(a) AOCB is *not* a cyclic quadrilateral. Why not?
(b) If reflex angle AOC = 200°, what is the size of angle ABC?

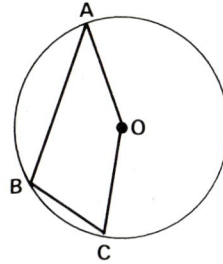

Fig. 18:15

*5 Try to draw the following polygons so that each corner lies on the circumference of a circle (draw the circles first).
(a) An obtuse-angled triangle.
(b) An isosceles triangle.
(c) An equilateral triangle.
(d) A square.
(e) A rectangle.
(f) A parallelogram.
(g) A rhombus.
(h) A kite.
(i) An isosceles trapezium.
(j) A non-isosceles (scalene) trapezium.

6 Write down the five cyclic quadrilaterals in Figure 13:9 (see Exercise 13B, For Discussion).

7 Both pairs of opposite angles of a certain cyclic quadrilateral are equal. What special kinds of quadrilateral could it be?

8 In Figure 18:16 calculate, with reasons:
(a) g and e if $f = 100°$
(b) g and e if $f = 120°$
(c) g and e if $f = 75°$.

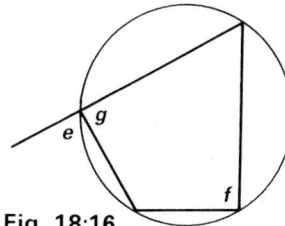

Fig. 18:16

You should have found that $e = f$ for any value of f. This is stated as:

The exterior angle of a cyclic quadrilateral is equal to the opposite interior angle.

9 Copy Figure 18:17, where AB = AC and ADE is a straight line. Prove that ABCD is a cyclic quadrilateral.

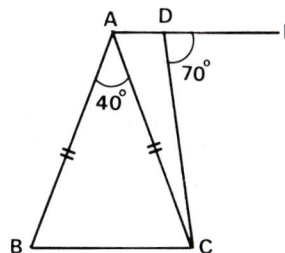

Fig. 18:17

10 In Figure 18:18 ABC is a straight line. Prove that ∠AOX = 2∠CBX.

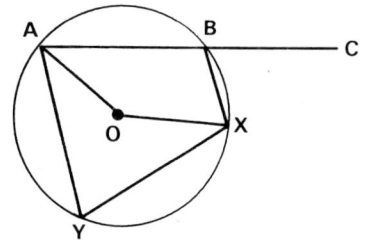

Fig. 18:18

11 For Figure 18:19:
(a) Calculate, giving reasons, angles *f* and *b*.
(b) Why must DA be parallel to CB?
(c) Show that DA is always parallel to CB for any value of ∠DAE (say *x*°).

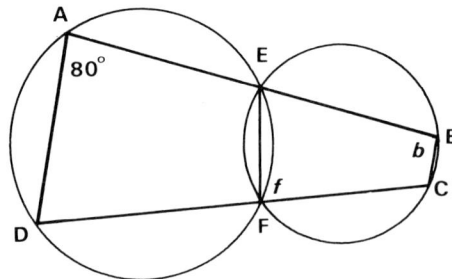

Fig. 18:19

12 Show that quadrilateral ABCD, with angles in the ratio A:B:C:D = 3:4:6:5, is a cyclic quadrilateral.

13 AB is the diameter of a circle. C and D are points on the circumference such that ∠CAB = 30° and DB bisects ∠ABC. Prove that CA bisects ∠DAB.

14 The side RS of cyclic quadrilateral PQRS is produced (made longer in the direction RS) to a point T. If PR = PQ prove that PS is the bisector of angle TSQ.

Reminders

$3a^2$ means $3 \times a^2$, so that if $a = 4$, $3a^2 \rightarrow 3 \times 16 = 48$.

$3a^2b^3 \times 2ab^4 \rightarrow 6a^3b^7$. You may use the rule 'Add the indices when multiplying powers of the same letter', or think of the terms written out in full:

$3a^2b^3 \times 2ab^4 \rightarrow 3 \times a \times a \times b \times b \times b \times 2 \times a \times b \times b \times b \times b$

3×2 is 6; the three a's multiply to give a^3; the seven b's multiply to give b^7.

1 If $a = 1$, $b = 3$, $c = 4$ and $h = 0$, find the value of:
 (a) $2a^3$ (b) $2b^2$ (c) a^3h^2 (d) ab^2h (e) $2b^2 - c^2$ (f) $b^3 - 4b^2h^3$.

2 Simplify:
 (a) $a^4 \times a^3$ (b) $b \times b^2$ (c) $2b \times 2b$ (d) $3a^2 \times 2a^3$ (e) $a \times b$ (f) $a^3 \times 4b^2$.

3 Simplify:
 (a) $4c^3 \times 3c^5$ (b) $3c^2 \times 5d^2$ (c) $2a^2 \times 3ab^2$ (d) $4m^3n^2 \times 3mn^3$.

4 **Example**   To simplify $4a^5b^3c \div 8a^2bc^3$.

$$\frac{4a^5b^3c}{8a^2bc^3} \rightarrow \frac{^1 4 \times \not a \times \not a \times a \times a \times a \times \not b \times b \times b \times \not c}{_2 8 \times \not a \times \not a \times \not b \times \not c \times c \times c}$$

Having cancelled as much as possible, this leaves $\dfrac{a^3b^2}{2c^2}$.

Or: Using the 'subtract indices' rule:

$$\frac{4a^5b^3c}{8a^2bc^3} \rightarrow \frac{1a^{5-2}b^{3-1}}{2c^{3-1}} \rightarrow \frac{a^3b^2}{2c^2}$$

Note: The c term is at the bottom in the answer because the bigger power of c (c^3) was at the bottom to start with.

Simplify:
 (a) $\dfrac{a^4}{a^3}$ (b) $\dfrac{a^8}{a^4}$ (c) $\dfrac{a^4}{a}$ (d) $\dfrac{a^3}{a^3}$ (e) $a^6 \div a^4$.

5 Simplify:
 (a) $a^6 \div a^6$ (b) $a^3 \div a^5$ (c) $a^2 \div a^6$ (d) $a \div a^6$.

*6 Simplify:
 (a) $\dfrac{a^5}{a^3}$ (b) $\dfrac{2a^6}{a}$ (c) $\dfrac{3a^3}{15a^2}$ (d) $\dfrac{24b^3c}{6bc}$ (e) $\dfrac{2a^2c^3}{3c^3d^2}$.

*7 Substitute the values $a = 2$, $b = 5$, $c = 4$ and $d = 3$ into your *answers* to question 6.

*8 **Example** $c^3(c^4 + 2) \rightarrow c^7 + 2c^3$

Multiply:
 (a) $x^2(x^3 + 3)$ (b) $x^3(x - 5)$ (c) $x^3(x^2 - x)$.

9 **The value of a^0**

$a^x \div a^x$ must equal 1, but using the subtraction of indices rule, $a^x \div a^x \rightarrow a^{x-x} = a^0$. Hence the fact that

$$a^0 = 1$$

Find the value of:
(a) 3^0 (b) 8^0 (c) c^0 (d) $4a^0$ (e) $3b^0$.

10 **Example** $(2a^2)^3 \rightarrow 2a^2 \times 2a^2 \times 2a^2 \rightarrow 8a^6$

Learn: $(x^m)^n \rightarrow x^{mn}$

Simplify:
(a) $(3a)^2$ (b) $(-a)^3$ (c) $(3ac)^3$ (d) $(-4)^3$ (e) $(-ab)^3$ (f) $(3a^2)^4$

(g) $(2a^2b)^2$ (h) $\left(\dfrac{1}{b^2}\right)^3$ (i) $\left(\dfrac{2a}{3b}\right)^3$ (j) $\left(\dfrac{a^2}{b^3}\right)^2$

11 $x^{\frac{1}{2}}$ is another way of writing $\sqrt{x}$ because $x^{\frac{1}{2}} \times x^{\frac{1}{2}} = x^{\frac{1}{2}+\frac{1}{2}} = x$.

Similarly $x^{\frac{1}{3}} \times x^{\frac{1}{3}} \times x^{\frac{1}{3}} = x$, so $x^{\frac{1}{3}}$ is the cube root of x, $\sqrt[3]{x}$.

Learn: $x^{\frac{1}{n}}$ **is the same as** $\sqrt[n]{x}$.

A fractional index gives a root.

Find:
(a) $9^{\frac{1}{2}}$ (b) $25^{\frac{1}{2}}$ (c) $27^{\frac{1}{3}}$ (d) $125^{\frac{1}{3}}$ (e) $16^{\frac{1}{4}}$.

Your calculator may mark its root key using the above notation. Look back to Chapter 4, page 28, for more information on this.

12 As 8^2 means '8 squared', and $8^{\frac{1}{3}}$ means 'the cube root of 8', it follows that $8^{\frac{2}{3}}$ means the cube root of 8 squared, that is, $2^2 = 4$.

Always find the root first, then the power; this keeps the numbers smaller.

Find:
(a) $27^{\frac{2}{3}}$ (b) $16^{\frac{3}{4}}$ (c) $4^{\frac{3}{2}}$ (d) $9^{\frac{3}{2}}$ (e) $8^{\frac{5}{3}}$ (f) $25^{1\frac{1}{2}}$ (g) $4^{2\frac{1}{2}}$.

13 By subtracting indices, $a^2 \div a^4 = a^{-2}$.

By cancelling, $a^2 \div a^4 \rightarrow \dfrac{a^2}{a^4} \rightarrow \dfrac{1}{a^2}$

Therefore $a^{-2} = \dfrac{1}{a^2}$.

Similarly $a^{-1} = \dfrac{1}{a}$ and $49^{-\frac{1}{2}} = \dfrac{1}{\sqrt{49}} = \frac{1}{7}$.

Learn: x^{-n} **is the same as** $\dfrac{1}{x^n}$.

A negative index gives 'one over'.

Find:
(a) 4^{-1} (b) 3^{-1} (c) 8^{-1} (d) 1^{-2} (e) 2^{-2}.

14 Simplify:
(a) $8^{-\frac{1}{3}}$ (b) $4^{-\frac{1}{2}}$ (c) $36^{\frac{1}{2}}$ (d) $36^{-\frac{1}{2}}$ (e) $25^{\frac{1}{2}}$.

15 Example   Evaluate $8^{-\frac{2}{3}}$.

Negative index $\Rightarrow$ one over. Hence $8^{-\frac{2}{3}} \to \dfrac{1}{8^{2/3}}$.

Index $\frac{1}{3} \Rightarrow$ cube root. Hence $\dfrac{1}{8^{2/3}} \to \dfrac{1}{2^2} \to \dfrac{1}{4}$.

Simplify:
(a) $27^{-\frac{2}{3}}$ (b) $16^{\frac{3}{4}}$ (c) $16^{-\frac{3}{4}}$ (d) $9^{-\frac{3}{2}}$ (e) $8^{\frac{5}{3}}$.

16 Simplify:
(a) $4^{-1\frac{1}{2}}$ (b) $16^{-1\frac{1}{2}}$ (c) $125^{-\frac{2}{3}}$ (d) 100^{-3} (e) $64^{\frac{2}{3}}$.

17 Simplify:
(a) $(3a)^{-2}$ (b) $3a^{-2}$ (c) $(2a)^{-1}$ (d) $2a^{-1} \times 2a^3$ (e) $(2a)^{-4}$
(f) $4^{1\frac{1}{2}}$ (g) $2a^{\frac{1}{2}} \times 3a^{\frac{1}{2}}$ (h) $\sqrt[3]{2^6}$ (i) $9^{-\frac{1}{2}}$ (j) $(36a^2)^{\frac{1}{2}}$
(k) $3^{-3} \times 3^4$ (l) $\sqrt{1\frac{9}{16}}$ (m) $\sqrt{6\frac{1}{4}}$ (n) 1^{-1} (o) $2^{\frac{1}{2}} \times 2^{\frac{3}{2}}$ (p) $0.04^{\frac{1}{2}}$
(q) 7^0 (r) $0.027^{\frac{2}{3}}$ (s) $3^a \times 3^{-a}$ (t) $8a \times (2a)^{-3}$.

●   You need to know . . .

● How to change a fraction to a percentage

Multiply the fraction by 100%.

Note that because $100\% = \frac{100}{100} = 1$ we do not increase the fraction when we multiply it by 100%.

Example $\dfrac{11}{15} \rightarrow \dfrac{11}{{}_3\cancel{15}} \times \overset{20}{\cancel{100}}\% \rightarrow \dfrac{220\%}{3} \rightarrow 73\tfrac{1}{3}\%$

● How to find a percentage of an amount

Example   To find 35% of £45.

35% is another way of writing $\dfrac{35}{100}$.

Hence 35% of £45 $\rightarrow \dfrac{35}{100} \times$ £45 $\rightarrow \dfrac{{}^7\cancel{35}}{{}_{20}\cancel{100}} \times$ £$\cancel{45}^{9} \rightarrow$ £15.75

● How to find one amount as a percentage of another

Write it as a fraction, then use the method of the first note above.

Example   To find 32 as a percentage of 128.

Write this as $\dfrac{32}{128}$, then $\dfrac{32}{128} \rightarrow \dfrac{32}{128} \times 100\% \rightarrow \dfrac{{}^8\cancel{32}^{1}}{{}_{32}\cancel{128}} \times 100\% \rightarrow 25\%$

Remember: One amount as a percentage of another amount is the first over the second times 100%.

● Percentage changes

A change can be an increase, a decrease, a profit, a loss, etc.

Example   To find the percentage loss if a book bought for £20 is re-sold for £18.

The change in the cost is £2.
The original cost was £20.
Hence the percentage loss is $\dfrac{2}{20} \times 100\% = 10\%$.

● To increase or decrease by a percentage

It is possible just to find the increase, then add it on, but a better method to use is:

To increase by $r\%$ multiply by $\dfrac{100 + r}{100}$.

To decrease by $r\%$ multiply by $\dfrac{100 - r}{100}$.

Examples   To increase by 12% you multiply by $\dfrac{100+12}{100} \rightarrow \dfrac{112}{100}$ or 1.12

To decrease by 12% you multiply by $\dfrac{100-12}{100} \rightarrow \dfrac{88}{100}$ or 0.88

● Using a calculator percentage key

Unfortunately different calculators do not always use the same method, but one of the following two methods will probably work.

To find 8% of £60. Key: 60 $\boxed{\times}$ 8 $\boxed{\%}$ **or** 60 $\boxed{\times}$ 8 $\boxed{\%}$ $\boxed{=}$

To increase £60 by 8%. Key: 60 $\boxed{+}$ 8 $\boxed{\%}$ **or** 60 $\boxed{+}$ 8 $\boxed{\%}$ $\boxed{=}$

To decrease £60 by 8%. Key: 60 $\boxed{-}$ 8 $\boxed{\%}$ **or** 60 $\boxed{-}$ 8 $\boxed{\%}$ $\boxed{=}$

Test yourself

1   (a) Express as a decimal fraction:
(i) 20%   (ii) 35%   (c) $12\frac{1}{2}$%.

(b) Express the above percentages as common fractions in their lowest terms.

2   Express as a percentage:
(a) $\frac{1}{4}$ (b) $\frac{4}{5}$ (c) $\frac{5}{8}$ (d) $\frac{1}{3}$ (e) 0.42   (f) 0.02   (g) 0.625

3   Find:
(a) 40% of £2   (b) 10% of £75   (c) 30% of 150 g   (d) 140% of 15 m
(e) 52% of £2.50   (f) $2\frac{1}{2}$% of £120.

4   Find:
(a) £6 as a percentage of £24.   (b) 9 as a percentage of 45
(c) 28 cm as a percentage of 35 cm   (d) 57 p as a percentage of 19 p
(e) 3.5 kg as a percentage of 2.8 kg.

5   (a) Increase 50 by 20%.   (b) Decrease 20 by 10%.
(c) Increase 50 kg by 25%.   (d) Decrease 96 m by 15%.

6   Find the profit or loss, saying which it is, if:
(a) an article costing £15 is then sold for £18
(b) an article costing £20 is then sold for £18.

7   A pupil is given 48 marks out of 60. What percentage is this?

8   Gunpowder is made of 75% nitre, 10% sulphur and 15% charcoal. How many grams of each are needed to make 500 g of gunpowder?

9   A man buys a carpet for £400 and sells it for £500. Find his profit percent.

10   Lois gives 30% of her wages to her mother. How much does she give if she earns £70?

11 Hardcheese school has 725 pupils, of whom 29 are absent on April 1st. What percentage is present?

12 My petrol tank holds 45 litres when full. How many litres are in the tank when it is 30% full?

13 Electromart allows 5% discount for cash. Find the reduced price of a television marked at £324.

---

14 Find $8\frac{1}{3}$% of 84 g.

15 What is $37\frac{1}{2}$% of 30 kg?

16 Joseph buys a cycle for £153 and sells it a year later for £136. Find the loss percent.

17 A wholesaler charges £2.25 for a can of paint which the retailer sells as a special offer for £2.52. What is the retailer's profit percent?

18 Potatoes are bought at £4.80 for 56 kg and sold at 16 p per kg. Find the percentage profit:
(a) reckoned on the cost price
(b) reckoned on the selling price.

19 Harvey invests £5000, and after one year is pleased to find that it has appreciated by 6%. In the next three years it appreciates by 10%, 4% and 10%.

Find:
(a) the value of his investment at the end of each year
(b) the profit percent on his original investment at the end of four years, correct to 2 significant figures.

Percentage changes (inverse calculations)

Sometimes you know the amount resulting from a percentage change and have to find the amount before the change took place. This would occur in a shop where the price given includes 15% VAT and the customer wants to know what the price was before VAT.

A shopkeeper once told me that the VAT on a £100 television was £15, because 15% of £100 was £15. He was wrong, because the VAT was 15% of the 'before-VAT' price. This is what he should have done:

To increase by 15% multiply by 1.15,
so 1.15 × before-VAT price = £100
→ before-VAT price = £100 ÷ 1.15 = £86.96
The VAT was therefore £13.04.

Percentage profit or loss in mathematics questions is, by tradition, assumed to be reckoned on the cost price, unless stated otherwise.

Example Selling price £28, profit 12%, find the cost price.

The temptation is to work out 12% of £28, then take this away from the £28, but this is not correct, for the 12% profit is reckoned on the cost price. The correct method is:

To increase by 12% multiply by $\frac{112}{100}$ or 1.12

Then 1.12 × cost price = £28 → cost price = £28 ÷ 1.12
$$= £25.$$

Example Selling price £21, loss 40%, find the cost price.

To reduce by 40% multiply by $\frac{100-40}{100}$ → $\frac{60}{100}$ or 0.60

Then 0.60 × cost price = £21 → cost price = £21 ÷ 0.60
$$= £35.$$

1 Find the cost price for the following.

	(a)	(b)	(c)	(d)	(e)
Selling price	£45	£21	£15	£8	£5.20
Profit/loss	10% loss	40% profit	50% profit	20% loss	4% profit

2 (a) Increase 15 by 25%.

(b) After an increase of 25% a number becomes 35. Find the number.

(c) Decrease 60 by 55%.

(d) What number becomes 18 when it is decreased by 20%?

(e) Find the cost price of an article sold for £15 at a profit of 25%.

(f) With VAT of 15% a vase costs £46. What did it cost before VAT?

3 A dealer buys a book for £4 and sells it at a profit of 36%. Find the selling price.

4 Ahmed spends 24% of his money and has 38 p left. How much had he at first?

5 Leah buys a bicycle for £15 and sells it for £18. Her sister, Rachel, buys a bicycle for £27 and sells it for £36. Who makes the greater profit percent and by how much?

6 A cruise cost £720 after a 20% reduction. What was the original cost?

7 Hashi owns a dress shop. He reckons his profit as a percentage of his cost price. If he makes 75% profit on a dress selling for £105, what did the dress cost him?

8 Roger sells a car for £686 and reckons that he has lost 30% of what it cost him. What was the original cost?

9 Heather is given a $12\frac{1}{2}\%$ increase in her salary of £3880. What is her new monthly salary?

10 Lizzie is given a 7% pay rise, bringing her pay up to £481.50. What was her pay before the rise?

11 A manufacturer is asked to produce an operating lever. He is allowed a tolerance of 2% on the specified length, giving 153 mm as the longest acceptable lever. What is the specified length?

12 A car-battery manufacturer makes up the electrolyte to contain 30% acid and 70% water by volume. How many litres of electrolyte can he make from 25 litres of acid?

13 Find the original value of a car if after losing $8\frac{1}{3}\%$ of its value it is worth £528.

14 Shelley sells a record for £2.40, thereby losing 4% on what she paid for it. At what price should she have sold it to *gain* 6%?

---

15 A manufacturer sells an article to a wholesaler, who then adds 25% to the price before selling it to a shopkeeper. The shopkeeper adds 35% to his cost price and sells the article to a customer for £5.40.

(a) What did the manufacturer charge the wholesaler?

(b) If the manufacturer is making 60% profit and the cost of making the article is made up of raw materials, labour, and overheads in the ratio 1 : 4 : 3 respectively, what is the cost of the raw material?

16 Collect examples of the uses of percentages in everyday life. Write a report on your findings.

Using your calculator

The Nile

For 4145 miles, from 8594 feet up in Burundi to the Mediterranean, the Nile brings life and spectacle to nine African countries. At Khartoum, 1850 miles from the sea, the White Nile is joined by the Blue Nile, a thunderous river that provides six-sevenths of the total water that reaches the sea.

Without the Nile, Egypt, a country of 390 000 square miles, would be a barren desert. Of Egypt's 44 million people, 97% live on the 4% of its land watered by the Nile. Cairo, with ten million people, the largest city in Africa, lies 4000 miles from the source of the Nile at the start of the delta, 150 miles wide when it reaches the sea. Half of Egypt's population live in this delta.

For thousands of years Egypt was at the mercy of the annual flooding of the Nile, but between 1960 and 1970, 41 000 workmen built the Aswan Dam, $2\frac{1}{2}$ miles wide and 364 feet high, at a cost of £560 million. Now the flood waters, held back to make Lake Nasser, are released gradually, generating 8000 million kilowatt-hours of electricity annually, 70% of Egypt's needs. Surplus water is piped up to 30 miles inland, reclaiming millions of acres of desert. In the first ten years of the dam's life, Egypt's food production rose by 35%, with a bonus of 30 000 tons of fish being caught annually in the lake.

Helped by the dam and strikes of oil, Egypt's economy is improving rapidly, though in 1983 its average per capita income was still only £226 per annum.

Use this information and your calculator to answer the following questions.

1 How far is it from the source of the White Nile to Khartoum?

2 What percentage of the Nile, when it reaches the sea, started as the White Nile?

3 What fraction of the circumference of the Earth is the Nile? (Earth's radius = 6380 km. Take 5 miles = 8 km.)

4 About how many Egyptians live on land not watered by the Nile?

5 About how many hectares are watered by the Nile? (1 hectare = 10 000 m². Take 5 miles = 8 km.)

6 What is the area of the Nile delta? (Assume it is an isosceles triangle.)

7 How many kilowatt-hours of electricity does the dam generate each day? (Answer in standard form.)

8 How many kilowatt-hours of electricity does Egypt need each year? (Answer in standard form.)

9 How many man-hours did it take to build the Aswan Dam? (Assume each man worked 8 hours a day, and that work continued 'round the clock'.)

10 What percentage of the total income of all Egypt's population did the dam cost?

11 What is the area of one face of the dam in m²? (8 km $\simeq$ 5 miles, and 1 m = 3.28 ft.)

21 Graphs: general parabola and equation solving

A The general parabola: $y = ax^2 + bx + c$

Curve sketching

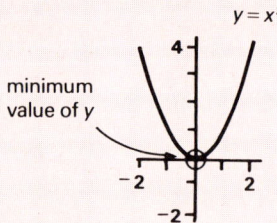

$y = x^2$

minimum value of y

Fig. 21:1

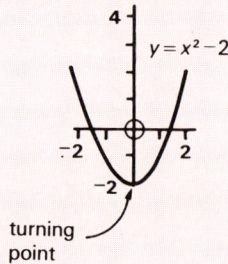

$y = x^2 - 2$

turning point

Fig. 21:2

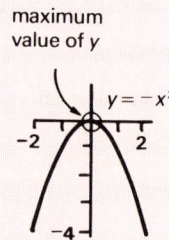

maximum value of y

$y = -x^2$

Fig. 21:3

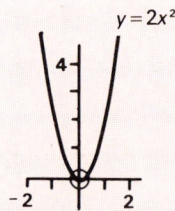

$y = 2x^2$

Fig. 21:4

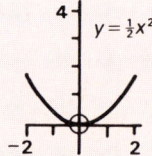

$y = \frac{1}{2}x^2$

Fig. 21:5

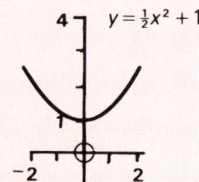

$y = \frac{1}{2}x^2 + 1$

Fig. 21:6

(a) All graphs of the family $y = ax^2 + c$ are symmetrical about the y-axis.

(b) The value of c gives the crossing point on the y-axis.

(c) The value of a affects the width of the curve and which way up it is: the higher the value of a the wider the curve becomes (using the same axes scales); if a is negative then the parabola has its turning point at the top (see Figure 21:3).

(d) Always draw a parabola with your wrist 'inside the curve', the way that your hand pivots naturally.

For Discussion

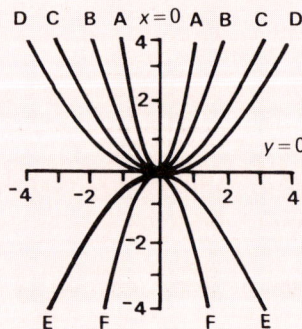

$y = 0$

Fig. 21:7

1 (a) Copy and complete the table for $y = 2x^2 + 3$.

x	-3	-2	-1	0	1	2	3
$2x^2$	18	8					
3	3	3					
y	21	11					

(b) Draw the graphs of $y = 2x^2$ and $y = 2x^2 + 3$ on one pair of axes. Scales: x from -3 to 3, 1 cm to 1 unit; y from 0 to 25, 1 cm to 5 units.

(c) Check the truth of the statement: 'The graph of $y = 2x^2 + 3$ is the same shape as the graph of $y = 2x^2$, but has its turning point at (0, 3) instead of at (0, 0)'.

2 (a) Copy and complete the table for $y = \frac{1}{2}x^2 - 1$:

x	-4	-3	-2	-1	0	1	2	3	4
$\frac{1}{2}x^2$	8	$4\frac{1}{2}$							
-1	-1	-1							
y	7	$3\frac{1}{2}$		$-\frac{1}{2}$					

(b) Draw the graphs of $y = \frac{1}{2}x^2$ and $y = \frac{1}{2}x^2 - 1$ on the same axes. Scales: x from -4 to 4, 1 cm to 1 unit; y from -1 to 8, 1 cm to 1 unit.

(c) Write a statement like the one in question 1(c) about this pair of graphs.

3 Copy and complete the following tables for the given equations. Do not draw any graphs yet.

(a) $y = x^2 + 2x + 1$

x	-4	-3	-2	-1	0	1	2
x^2	16						
$2x$	-8						
1	1						
y	9	4		0			

(b) $y = x^2 - 2x + 2$

x	-2	-1	0	1	2	3	4
x^2	4						
$-2x$	4						
2	2						
y	10						

(c) $y = x^2 - 3x + 1$

x	0	$\frac{1}{2}$	1	$1\frac{1}{2}$	2	$2\frac{1}{2}$	3
x^2				$2\frac{1}{4}$			
$-3x$				$-4\frac{1}{2}$			
1				1			
y		$-\frac{1}{4}$		$-1\frac{1}{4}$			

(d) $y = 2x^2 + x + 1$

x	-2	-1	0	1	2
$2x^2$	8				
x	-2				
1	1				
y	7				

(e) $y = 1 + 2x - x^2$

x	-2	-1	0	1	2	3
1	1					
$2x$	-4					
$-x^2$	-4					
y	-7					

*4 Draw, very carefully, the graphs for the equations in question 3, using the scales given below. Make sure that your parabolas are smooth curves, with no sharp corners, bumpy bits, or 'feathers'!

Remember to label each graph with its equation.

Graph	*x*-axis	*y*-axis
(a)	-4 to 2, 1 cm to 1 unit	0 to 10, 1 cm to 2 units
(b)	-2 to 4, 1 cm to 1 unit	0 to 10, 1 cm to 2 units
(c)	0 to 3, 2 cm to 1 unit	-2 to 1, 2 cm to 1 unit
(d)	-2 to 2, 2 cm to 1 unit	0 to 12, 1 cm to 2 units

Note: the lowest point is at $\left(-\frac{1}{4}, \frac{7}{8}\right)$.

(e)	-2 to 3, 1 cm to 1 unit	-7 to 2, 1 cm to 1 unit

5 **Example** The axis of symmetry of the parabola $y = x^2 + 2x + 1$ in question 3(a) is $x = -1$, because the y-values are symmetrical about this x-value.

State the equations of the axes of symmetry for the other graphs in question 3. (See the note in question 4 about the minimum value of y for graph (d).)

6 (a) Draw up a table for $y = 2x^2 - x - 10$ for the integral values of x from -3 to 4.

  (b) Using axes, x from -3 to 4, 1 cm to 1 unit; y from -12 to 20, 1 cm to 2 units, draw the graph of $y = 2x^2 - x - 10$.

  (c) State the equation of the axis of symmetry of your graph.

  (d) State the lowest possible value of y (its minimum value) if $y = 2x^2 - x - 10$. At what value of x does this occur?

7 **Example** Figure 21:8 shows the graph of the parabola $y = x^2 + 2x + 1$.

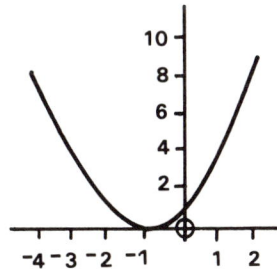

Fig. 21:8

At the point $(-1, 0)$ we know that $x = -1$ and $y = 0$.

As $(-1, 0)$ is a point on the parabola $y = x^2 + 2x + 1$ then when we substitute $x = -1$ into the equation the result must be 0.

We say that $x = -1$ is a solution of the equation $0 = x^2 + 2x + 1$ (usually written $x^2 + 2x + 1 = 0$).

  (a) Draw your own copy of the parabola $y = x^2 + 2x + 1$, using the table you drew up in question 3(a) and the scales given in question 4. Remember to label the curve.

  (b) Draw on your grid the line $y = 1$. Find the values of x at which this line crosses the parabola.

  (c) Your answers to part (b) are the solutions of the equation $1 = x^2 + 2x + 1$ (or $x^2 + 2x = 0$). Try to explain briefly why this is

  (d) Draw on your grid the line $y = 4$. At what values of x does $y = 4$ cross $y = x^2 + 2x + 1$? Show why these values of x make the equation $x^2 + 2x = 3$ true.

  (e) Draw on your grid the line $y = 6$. State, as accurately as you can, the values of x where $y = 6$ crosses $y = x^2 + 2x + 1$. State also the equation solved by these values of x.

  (f) Repeat part (e) for the line $y = 8$.

  (g) Draw on your grid the line $y = -1$. What must be true about the equation $x^2 + 2x = -2$?

8 (a) Draw the graph of $y = x^2 - 2x + 2$ (see question 3(b) and question 4(b)).

  (b) Draw the following lines on your grid. For each line state the values of x where it crosses the parabola and state the equation that you have solved.
  (i) $y = 1$ (ii) $y = 2$ (iii) $y = 5$ (iv) $y = 4$ (v) $y = 8$

9 State the equation solved by the x-co-ordinates of the crossing points of the following pairs of graphs. Do not draw the graphs.
(a) $y = 1$ and $y = x^2 - 3x + 2$ (b) $y = -1$ and $y = 2x^2 + 3x$
(c) $y = 2$ and $y = x^2 - x - 1$ (d) $y = 8$ and $y = 3x^2 - 2x + 7$
(e) $y = -2$ and $y = 2x^2 - 3x - 2$

---

10 Draw five graphs of $y = x^2$ for values of x from -3 to 3 using the following five scales:
(a) x, 1 cm to 1 unit; y, 1 cm to 1 unit
(b) x, 1 cm to 1 unit; y, 1 cm to 2 units
(c) x, 1 cm to 1 unit; y, 2 cm to 1 unit
(d) x, 2 cm to 1 unit; y, 1 cm to 1 unit
(e) x, 2 cm to 1 unit; y, 1 cm to 2 units

11 If all the axes in question 10 were re-numbered to 1 cm to 1 unit, without changing the drawn graphs, what would the equations of the graphs become?

B Using parabolas to solve quadratic equations

You should have completed Exercise 21A up to question 9 before starting this exercise.

A **quadratic equation** is one involving x^2. A very simple quadratic equation is $x^2 = 1$. Unlike the equations you have met before, a quadratic equation usually has two solutions. Both $x = 1$ and $x = -1$ make $x^2 = 1$ true.

Example Figure 21:9 shows the graph of $y = x^2 - 3x + 1$.

To solve the equation $x^2 - 3x + 1 = 0$ (or $x^2 - 3x = -1$).

The parabola's equation, $y = x^2 - 3x + 1$, becomes the equation we have to solve, $x^2 - 3x + 1 = 0$, when $y = 0$. Therefore we find the solutions where the parabola crosses the line $y = 0$ (the x-axis). They are $x = 2.6$ and $x = 0.38$ approximately. (You often do not have exact answers to a quadratic equation. Sometimes there are no solutions e.g. how can $x^2 + 1 = 0$?)

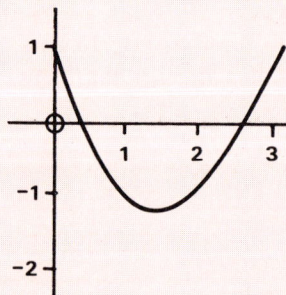

Fig. 21:9

Example To solve $x^2 - 3x + 2 = 0$ using the graph of $y = x^2 - 3x + 1$ (Figure 21:9).

First we change the left-hand side of the given equation to make it the same as the graph equation; that is, we have to change $x^2 - 3x + 2$ into $x^2 - 3x + 1$. We do this by subtracting 1:

$$x^2 - 3x + 2 = 0 \xrightarrow{-1 \text{ from both sides}} x^2 - 3x + 1 = -1.$$

Now the solutions may be read where $y = -1$.

Answer: $x = 1$ and $x = 2$.

Check that $x^2 - 3x + 2 = 0$ when $x = 1$ and when $x = 2$.

1 Copy and complete:

To solve $x^2 - 3x = 0$ using the graph of $y = x^2 - 3x + 1$.
Change $x^2 - 3x$ into $x^2 - 3x + 1$ by adding 1.
Then $x^2 - 3x = 0 \xrightarrow{+1 \text{ to both sides}} \ldots\ldots\ldots\ldots$
Read the solutions where $y = \ldots$ crosses the parabola.
Answer: $x = \ldots$ and $x = \ldots$

2 If you had drawn the graph of $y = 2x^2 + 8x - 5$, state the equation of the straight line you need to draw to solve:
(a) $2x^2 + 8x - 4 = 0$ (b) $2x^2 + 8x + 1 = 0$ (c) $2x^2 + 8x - 1 = 0$
(d) $2x^2 + 8x + 3 = 0$ (e) $2x^2 + 8x - 5 = 0$ (f) $2x^2 + 8x = 0$.

3 Draw the graph of $y = 2x^2 + 8x - 5$ for values of x from -5 to 1, using an x-axis scale of 2 cm to 1 unit. Use your graph to solve the equations given in question 2.

4 Draw on one grid, for values of x from -4 to 4, the graphs of $y = x^2$ and $y = -3x + 4$. State the equation solved by the crossing points of these graphs, and its solutions.

---

5 State the equation solved where $y = x^2$ crosses:
(a) $y = 2x + 1$ (b) $y = -2x + 1$ (c) $y = 3x - 2$
(d) $y = -2x - 4$ (e) $y = \frac{1}{2}x + 3$ (f) $y = \frac{1}{3}x - 1$.

6 Draw the graphs given in question 5 and hence solve the equations.

7 You can easily program a computer to give you the points to plot for a given equation of the type $y = ax^2 + bx + c$.

```
10 REM 'PLOTME'
20 PRINT "Points to plot for parabola of form y = ax² + bx + c."
30 PRINT "Type value of a."
40 INPUT A
50 PRINT "Type value of b."
60 INPUT B
70 PRINT "Type value of c."
80 INPUT C
90 PRINT "Type smallest value for x."      (Continued on next page)
```

```
100 INPUT SX
110 PRINT "Type largest value for x."
120 INPUT LX
130 PRINT "Type x interval (e.g. 1 for −3, −2, −1, etc.)"
140 INPUT I
150 CLS (Clear screen)
160 PRINT "y =ˇ";A;"x↑2ˇ+ˇ";B;"xˇ+ˇ";C
170 FOR X = SX TO LX STEP I
180 PRINT "Plot (ˇ";X;"ˇ,ˇ";A*X*X + B*X + C;"ˇ)"
190 NEXT X
```

22 Geometry: review (proofs)

Reminders

Quadrilaterals

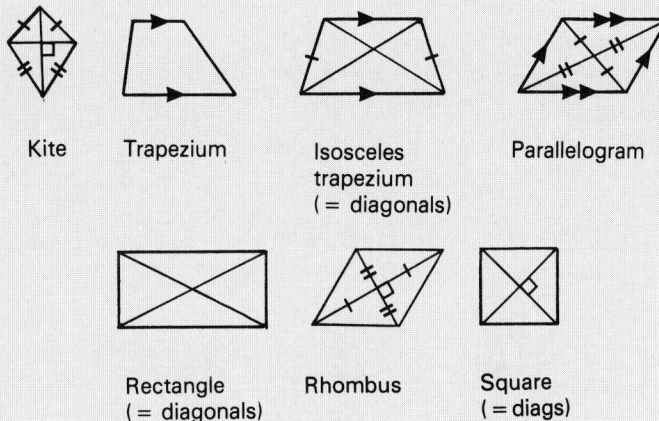

Kite   Trapezium   Isosceles trapezium (= diagonals)   Parallelogram

Rectangle (= diagonals)   Rhombus   Square (= diags)

Fig. 22:1

Properties can be observed from an accurate sketch.

Important facts:

A parallelogram has no line of symmetry.

The diagonals of a kite, a rhombus, and a square cross at right angles.

The angle sum of all quadrilaterals is 360°.

Polygons

polygon: many-sided

pentagon: 5 sides   octagon: 8 sides

hexagon: 6 sides   nonagon: 9 sides

heptagon: 7 sides   decagon: 10 sides

The exterior angles of all polygons total 360° (Figure 22:2).

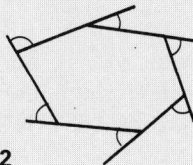

Fig. 22:2

The interior angles of an n-sided polygon total $(n - 2) \times 180°$ (Figure 22:3).

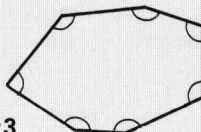

Fig. 22:3

22

Before working the exercise you may find it helpful to refer to the notes in Chapter 5.

1 In Figure 22:4, PQ // RS // TU, and QT is a straight line.

(a) What kind of angle is:
   (i) *a*   (ii) *b*   (iii) *c*?

(b) Give the reason why:
   (i) *a* = *e*   (ii) *a* = *g*   (iii) *a* + *b* = 180°   (iv) *b* + *d* = 180°   (v) *a* = *d*.

(c) (i) If *a* = 75° find *f*.   (ii) If *e* = 80° find *d*.   (iii) If *g* = 62° find *e*.
   (iv) If *f* = 142° find *g*.   (v) If *b* = 113° find *f*.   (vi) If *f* = 125° find *c*.

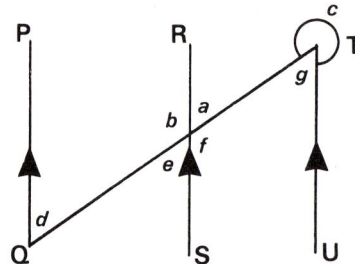

Fig. 22:4

2 In Figure 22:5 calculate *x* and hence the size of each angle.

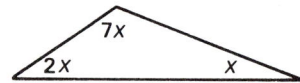

Fig. 22:5

3 In Figure 22:6 AB = AC. Calculate exterior angle BCE.

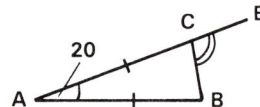

Fig. 22:6

4 (a) In Figure 22:7, M is the midpoint of AB, and MN // BC. Draw an enlarged copy of the figure, using any shape triangle. Check that AN = NC and that MN = $\frac{1}{2}$BC.

(b) Repeat part (a) with another shape of triangle.

(c) The results of part (a) are true for any triangle. This is called The Midpoint Theorem.

(d) Check by drawing that:
   'The line joining the midpoints of two sides of a triangle is parallel to, and a half of, the third side.'

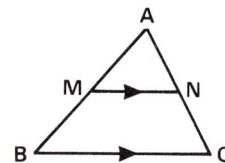

Fig. 22:7

5 In Figure 22:8, X is any point on AB and XY // BC.

It can be proved that $\dfrac{AX}{XB} = \dfrac{AY}{YC}$.

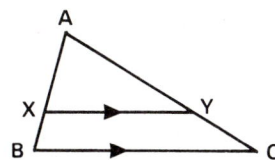

Fig. 22:8

(a) If AX = 5 cm, XB = 3 cm, and YC = 4 cm, calculate AY.

(b) If AB = 15 cm and AY:YC = 3:2, find AX and XB.

(c) Prove that △s AXY and ABC are similar, then write the ratio-of-the-sides equation (see Exercise 5B).

(d) If AX = 4 cm, XB = 3 cm, and XY = 8 cm, sketch △ABC then calculate BC using the equation in part (c).

6 Name the three special quadrilaterals in which the diagonals cross at 90°.

7 Name the three special quadrilaterals in which the diagonals are equal.

8 Calculate the size of an interior angle in a regular nonagon.

9 How many sides has a polygon if its interior angle sum is 1620°?

10 Four of the exterior angles of an irregular pentagon are each 80°. Calculate:
(a) the fifth exterior angle   (b) the five interior angles.

11 In which regular polygon are the interior and exterior angles in the ratio 4:1?

12 In an octagon how many diagonals can be drawn:
(a) from one vertex   (b) altogether?

*13 Copy the diagrams in Figure 22:9 and calculate the size of each lettered angle.

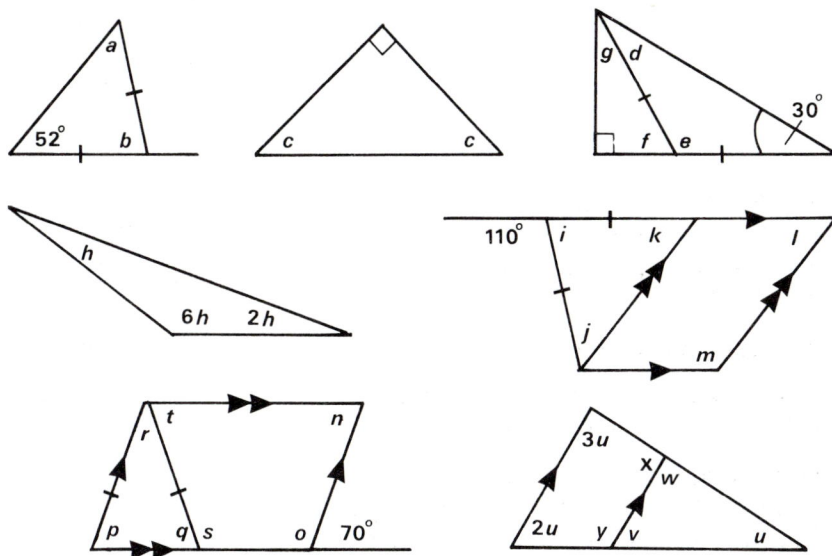

Fig. 22:9

*14 Make enlarged copies of the diagrams in Figure 22:10. Calculate the sizes of all angles less than 180° and write your answers on your diagrams.

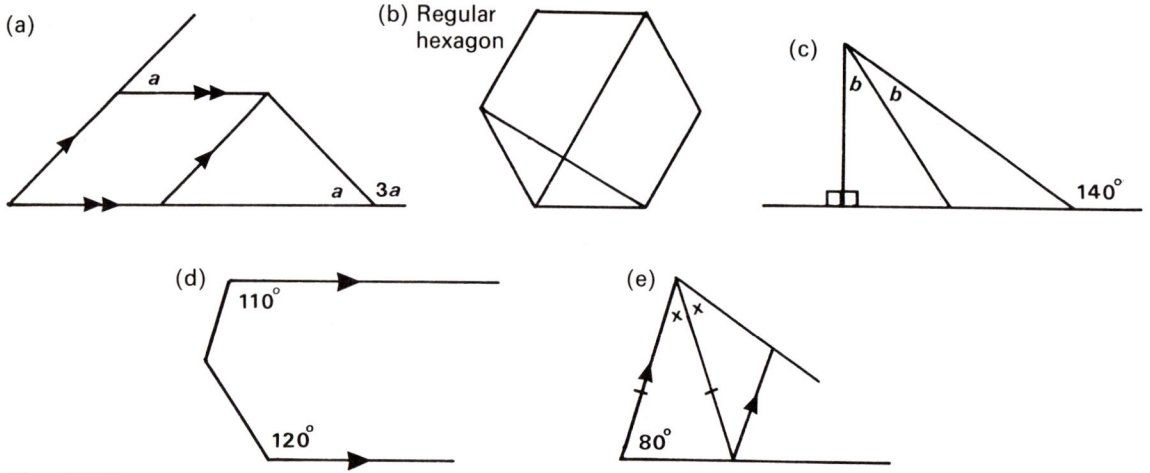

(a)

(b) Regular hexagon

(c)

(d)

(e)

Fig. 22:10

15 Sketch a regular pentagon PQRST. Join QT and QS. Calculate and write on your sketch the sizes of the angles in triangles QRS, PQT and QTS.

16 In a trapezium ABCD, AB // DC and AB = AD. Prove that BD bisects ∠ADC.

17 In Figure 22:11 prove that ∠ADC = 90°.

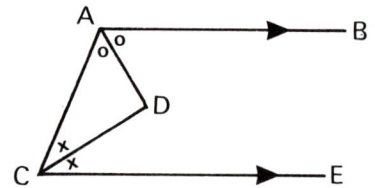

Fig. 22:11

18 In Figure 22:12 prove by congruent triangles that BQ = PC.

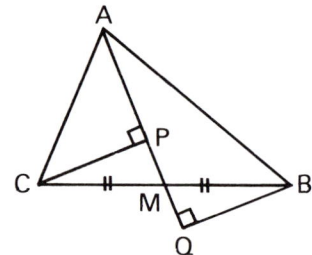

Fig. 22:12

19 In Figure 22:13, prove that BC = CD.

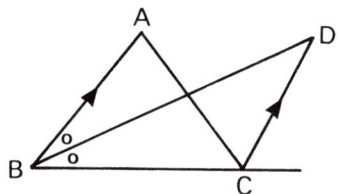

Fig. 22:13

20 Prove by congruent triangles that a quadrilateral with one pair of sides both equal and parallel must be a parallelogram.

21 ABCD is a rectangle. An isosceles triangle BCE is drawn outside the rectangle, with BE = CE. Prove that AE = DE.

22 The diagonals HJ and IK of parallelogram HIJK cross at M. Prove that the area of △HMK equals the area of △JMI. (Note: This is often just written: Prove △HMK = △JMI.)

23 In △DEF, M is the midpoint of EF and N is the midpoint of DM. Prove that:
(a) △DEM = △DMF   (b) △DEN = $\frac{1}{2}$△DMF.

24 K, L and M are the midpoints of the sides of any triangle PQR. Prove that:
(a) △KLM is similar to △PQR   (b) △KLM = $\frac{1}{4}$△PQR.

25 (a) Prove that the area of a kite is half the product of its diagonals.

(b) GHIJ is a kite with GH = GJ. A, B, C and D are the midpoints of GH, HI, IJ and JG respectively. Prove that:
(i) ABCD is a rectangle   (ii) area ABCD = $\frac{1}{2}$ area GHIJ.

26 S is a point on side PR of △PQR such that ∠RSQ = ∠PQR. If ∠P = 50° find ∠SQR.

27 ABCD is a parallelogram. AB is produced to F such that AB = BF. AD is produced to E such that AD = DE. Prove that EF passes through C. (Hint: Join E and F to C, then prove that ECF is a straight line.)

28 In △ABC, ∠A = 50°. The bisectors of angles B and C meet at D. Calculate ∠BDC.

29 ABCD is a parallelogram in which ∠BAC = 90°. BC is produced to E such that BC = 2CE and ∠DEC = 90°. Prove that ∠B = 45°.

---

30 Write a computer program that uses Pythagoras' Theorem to state whether a triangle whose three sides are given is acute-, obtuse-, or right-angled.

● You need to know . . .

● The three statistical averages

Mean
The total score distributed equally to each item. The mean of 2, 7 and 9 is the total (18) divided by 3, giving 6.

Mode
The item that occurs most frequently. There may be several modes, or no mode.

Median
The middle score of an odd number of items. For an even number of items it is usually taken as the mean of the two middle scores, although it could be anywhere between them.

● Frequency tables

These are used to simplify the arithmetic with a large amount of data. (Processing such large amounts takes a long time, so examples in books usually apply the method to a small amount of data!)

Example To find the mean of £16, £16, £17, £17, £17, £18, £18, £18, £18, £19, £19, £19, £20.

Score (x)	Frequency (f)	Total ($f \times x$)
£16	2	£32
£17	3	£51
£18	4	£72
£19	3	£57
£20	1	£20
TOTALS	13	£232

Mean = $\dfrac{£232}{13} \simeq £18$

Mode = £18 (highest frequency)

Median = £18 (score of the 7th item, which is clearly in the £18 frequency: 2 + 3 = 5;
2 + 3 + 4 = 9)

Test yourself

1 Find (a) the mean   (b) the mode(s)   (c) the median:
   (i) 7, 2, 4, 3, 4   (ii) 37, 32, 34, 33, 34 (look at (a))   (iii) 10, 16, 24, 32, 15, 11
   (iv) 109, 116, 124, 132, 116, 109, 113   (v) 7, −4, 12, −9, −4, 0, −2   (vi) 1, 7, 8, 4, 8, 8

2   Find the mean of:
(a) 13 years, 15 years, 12 years
(b) 8 min, 12 min, 11 min
(c) 12 years 6 months, 8 years 9 months, 14 years 7 months, 16 years 2 months.

*3   The midday Celsius temperatures during a week in August were: 24, 25, 28, 26, 22, 20 and 23 °C.

What was the mean average for the week?

*4   The wages of 8 people in a firm are: £108, £112, £96, £83, £130, £112, £118, and £121.

(a) What is:
(i) the mean wage   (ii) the median wage?

(b) If the managing director, earning £335, is included, what would be:
(i) the new mean   (ii) the new median?

*5   Joanna's monthly earnings for a year were: £540, £568, £539, £572, £572, £568, £570, £581, £578, £579, £572, and £565.

What was her mean average monthly earning?

*6   The ages of a family of five are:
40 years 5 months, 37 years 3 months, 16 years 9 months, 14 years 6 months, and 11 years 11 months.

What is the mean age of the family?

*7   In a race the time recorded by members of a team of six were:
13 min 42 s,   13 min 50 s,   13 min 57 s,   14 min 3 s,   14 min 29 s,   14 min 59 s.

What was:
(a) the total time taken by the team
(b) the mean time?

*8   The price of a Betex washing machine in six different shops was: £258, £236, £288, £274, £252, and £264.

(a) What was the mean price?
(b) How much cheaper than the mean was the cheapest?
(c) How much more expensive than the mean was the most expensive?

9   During a January the midday temperatures in °C were:
3, 7, 3, 11, 10, 3, 0, 0, −3, −8, −7, −5, −8, −8, −3, 2, 5, 6, 9, 7, 5, 5, 5, 5, 8, 6, 5, 1, 10, 9, 10.

What was:
(a) the mean temperature   (b) the modal temperature   (c) the median temperature?

10   The mean average age of four girls is 14 years 7 months. If three of the girls have ages of 14 years 11 months, 14 years 4 months, and 16 years 6 months, what is the age of the fourth girl?

11 The marks scored in a mental arithmetic test by a class of 30 pupils were:

Marks	0	1	2	3	4	5	6	7	8	9	10
No. of children	0	3	0	1	2	4	5	6	5	3	1

What was:
(a) the modal mark   (b) the median mark   (c) the mean mark?

12 The members of a school club took part in a sponsored walk and the following record was made of their sponsors.

Number of sponsors	6	9	11	12	15	16	17	19
Number of walkers	2	3	14	8	16	2	3	2

Find:
(a) the median number of sponsors   (b) the modal number of sponsors
(c) the mean number of sponsors.

13 By grouping the following marks in ranges 0–9, 10–19, 20–29, 30–39, and 40–50, calculate an approximation to the mean mark of the 100 pupils represented.

40, 43, 35, 33, 43, 45, 30, 40, 38, 46, 33, 40, 43, 34, 34, 40, 47, 41, 34, 23, 41, 35, 31, 44, 28, 42, 45, 43, 31, 37, 37, 37, 29, 46, 36, 19, 43, 37, 32, 31, 34, 42, 24, 34, 33, 41, 47, 36, 26, 43, 29, 22, 36, 26, 32, 29, 18, 36, 40, 33, 30, 24, 32, 29, 22, 26, 28, 36, 30, 25, 31, 18, 37, 26, 35, 35, 32, 24, 32, 36, 20, 39, 35, 35, 31, 16, 35, 44, 38, 36, 18, 23, 16, 36, 21, 23, 32, 32, 17, 30.

14 A cinema manager had to sell an average of 240 tickets a day to 'break even'. On the first six days of a week he sold 234, 180, 202, 233, 320, and 330 tickets.

(a) How many tickets had he to sell during the seventh day to avoid making a loss?

(b) If he sold 372 tickets on the seventh day and the tickets cost on average £2 each, how much profit did he make in the week?

15 During a twelve-month period Adrian used a rain gauge which enabled him to record the following monthly rainfalls (in mm):
84, 111, 49, 70, 28, 56, 98, 26, 104, 27, 30, 31.

(a) What was the mean rainfall per month?

(b) If the rainfall for the next five months was 116, 109, 41, 115, and 60 mm, what would the average have to be for the remaining seven months so that the twelve-month mean rainfall was the same as the previous year?

16 In a class examination 16 boys averaged 68% and 14 girls averaged 71%. What was the average mark for the 30 pupils?

17 The average age of six men is 24 years 5 months, and of five of them is 23 years 11 months. How old is the sixth man?

18 In a factory, parts were tested in batches of 1000, one batch per day. During February the number of rejects were:

Week One   8, 6, 4, 7, 4, 5, 2

Week Two   11, 0, 2, 3, 0, 3, 2

Week Three   10, 3, 6, 5, 4, 2, 3

Week Four   14, 6, 3, 2, 4, 0, 5

(a) What was the mean reject rate per batch:
   (i) each week   (ii) for the month?

(b) The firm expected a reject rate of not more than 0.5%. In which week did the reject rate surpass this allowance and by what percentage?

19 Design a computer program to give the mean, median and mode for sets of data. You might also illustrate the data with a diagram.

Modal class; mean from grouped data

For Discussion

One
Form 4B's exam marks:
51, 72, 37, 72, 39, 79, 38, 32, 68, 57, 34, 58, 62, 49, 47, 78, 61, 57, 63, 69, 48, 58, 42, 68, 52, 46, 63, 52, 57, 48.

Class	Tally	Frequency (f)	Middle (x)	Totals (fx)
30–39				
40–49				
50–59				
60–69				
70–79				

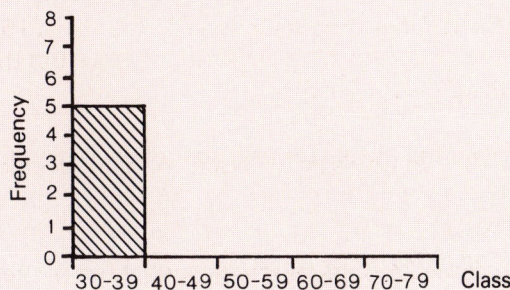

Fig. 23:1 **Bar-chart for 4B's exam**

23

Two

The class which has the highest frequency is called the **modal class**. As with the mode, there can be more than one modal class. In the example above the modal class is 50–59.

Three

To calculate the mean from grouped data you assume that each frequency scores the mid-value (middle) of the range. This mid-value is the mean of the upper and lower limits of the range, e.g. the middle of 0–9 is $\frac{0+9}{2} = 4.5$, and the middle of 31–35 is $\frac{31+35}{2} = 33$.

What is the middle of:
(a) 50–59   (b) 62–70   (c) 10–14?

Example

Class	Frequency (f)	Middle (x)	Totals (fx)
30–39	23	34.5	793.5
40–49	55	44.5	2447.5
50–59	32	55	1760
	110		5001

Mean $\simeq \frac{5001}{110} \simeq 45$

1 Calculate the mean of the data in the table.

Class	Frequency (f)	Middle (x)	Totals (fx)
0–9	5		
10–19	6		
20–29	8		
30–39	10		
40–49	7		
50–60	4		

2 The following data shows the absentees at a school each day during an 80-day term.

16, 29, 41, 6, 24, 31, 19, 46, 30, 27, 22, 35, 34, 36, 11, 25, 38, 45, 26, 39, 36, 52, 12, 51, 24, 35, 47, 53, 26, 4, 37, 32, 44, 31, 15, 43, 20, 31, 29, 39, 22, 38, 42, 32, 21, 38, 37, 34, 18, 33, 36, 34, 16, 42, 30, 28, 30, 37, 32, 20, 37, 27, 56, 43, 17, 57, 46, 33, 54, 47, 17, 39, 41, 48, 28, 35, 13, 31, 33, 21.

144

(a) Using class intervals of 0–9, 10–19, . . . , 50–59, make a frequency distribution table for this data.

(b) Represent the data by a bar-chart.

(c) Indicate the modal class on your bar-chart.

3 (a) Find the mean for the following tables of data.

(i)

Class	0–4	5–9	10–14	15–19	20–25
Frequency	3	5	6	4	2

(ii)

Class	1–5	6–10	11–15	16–20	21–25
Frequency	2	4	6	5	3

(iii)

Class	10–19	20–29	30–39	40–49	50–59	60–69	70–79	80–89
Frequency	2	10	15	25	22	18	5	3

(iv)

Class	5–9	10–14	15–19
Frequency	30	32	28

(b) Draw a pie-chart to represent the data in table (i).

4 An approximate modal value can be found by considering the ratio of the scores above and below the modal class, although the result is of very dubious value. It can best be done from a bar-chart. See Figure 23:2.

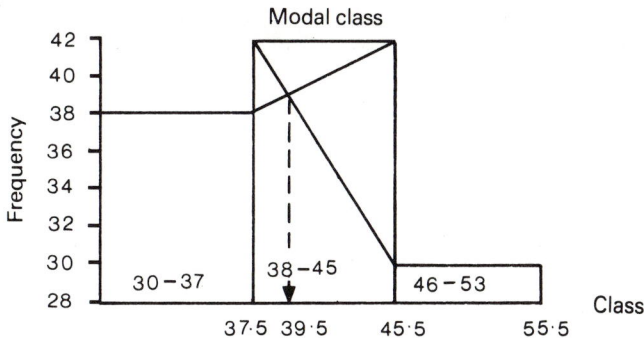

Fig. 23:2

An approximate value for the mode is 39.5

(a) Use the bar-chart drawn in the For Discussion to estimate the mode for the data.

(b) Estimate the modes from the bar-charts for question 3.

5 For the data in question 2 find:
(a) an estimate for the mean   (b) the true mode   (c) the true mean
(d) the median.

$$o = a \times \tan \theta \qquad o = h \times \sin \theta \qquad a = h \times \cos \theta$$

$$\frac{o}{a} = \tan \theta \qquad\qquad \frac{o}{h} = \sin \theta \qquad\qquad \frac{a}{h} = \cos \theta$$

One ancient teacher of history swore at his class!

Examples

Finding sides

$x = 5 \times \tan 54°$
$\simeq 6.88$

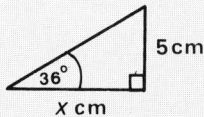

$x = 5 \times \sin 40°$
$\simeq 3.21$

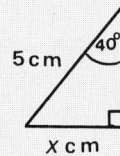

$x = 5 \times \cos 25°$
or $x = 5 \times \sin 65°$
$\simeq 4.53$

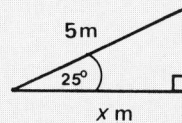

Sample key sequence: 5 ☒ 54 TAN =

Finding angles

$\frac{7}{6} = \tan \theta$

$\theta \simeq 49.4°$

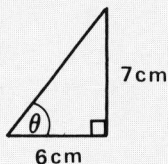

$\frac{7}{9} = \sin \theta$

$\theta = 51.1°$

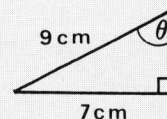

$\frac{7}{9} = \cos \theta$

or $\frac{7}{9} = \sin \alpha \rightarrow \alpha \simeq 51.1°$

$\theta \simeq 38.9°$

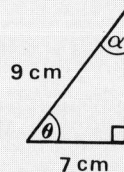

Sample key sequence: 7 ÷ 6 ARCTAN =

Note: ARCTAN may be INV TAN or TAN⁻¹.

In all questions give sides correct to 3 significant figures and angles to the nearest 0.1°.

1   Find the value of x or θ for each triangle in Figure 24:7.

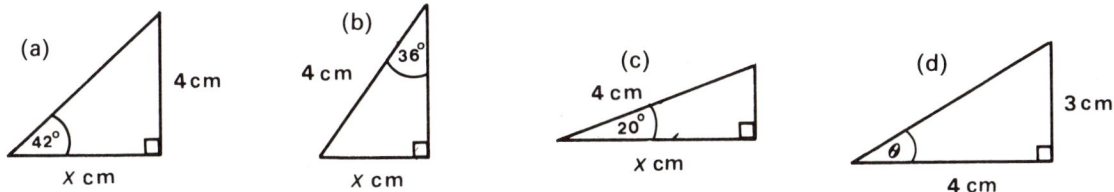

Fig. 24:7

2   Find the value of x or θ for each triangle in Figure 24:8.

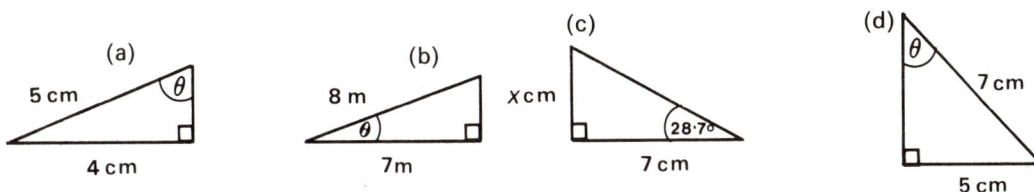

Fig. 24:8

3   Find the value of x or θ for each triangle in Figure 24:9.

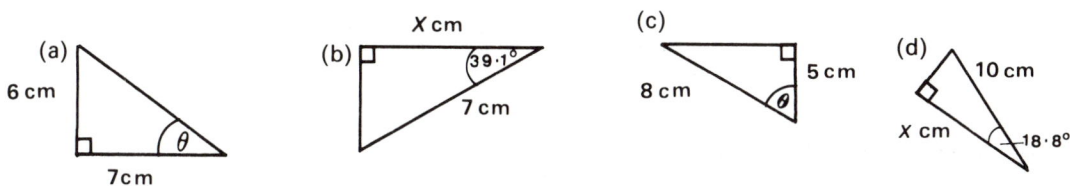

Fig. 24:9

4   Find the value of x or θ for each triangle in Figure 24:10.

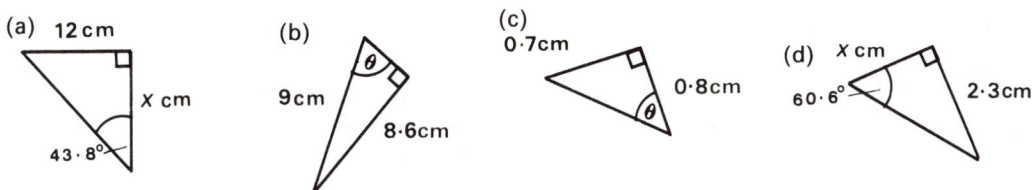

Fig. 24:10

*5   Sketch (not accurately) a triangle for each of the following. Give answers correct to 3 s.f. or 0.1°.

(a) ΔABC, ∠B = 90°; ∠A = 63.2°; AB = 4 cm.
    Calculate BC.

(b) ΔHIG, ∠G = 90°; HG = 4.5 cm; GI = 2 cm.
    Calculate ∠H.

(c) ΔJKL, ∠L = 90°; ∠J = 32.2°, JK = 11 cm.
    Calculate JL.

(d) △MNO, ∠M = 90°, ∠N = 32.6°, NO = 4.1 m.
Calculate MO.

(e) △PQR, ∠Q = 90°, PQ = 5 cm, PR = 7 cm.
Calculate ∠P.

(f) △VWX, ∠W = 90°, WX = 6 mm, ∠V = 18°.
Calculate VW.

*6   Use the angle sum of a triangle, Pythagoras' Theorem, or trigonometry, to find the remaining sides and angles in the triangles for question 5.

7   Copy each diagram in Figure 24:11, then calculate all sides, correct to 3 s.f., and all angles, correct to 0.1°.

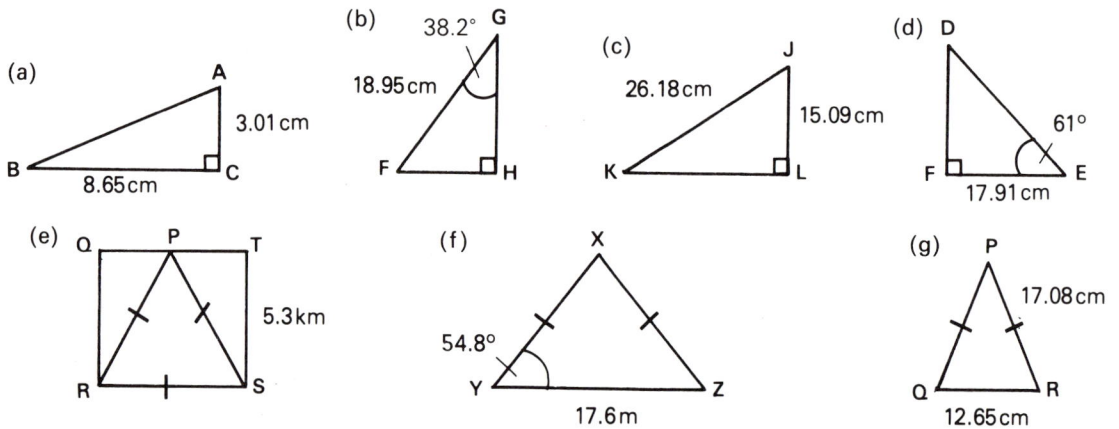

(a)

A
3.01 cm
B
8.65 cm
C

(b)
38.2° G
18.95 cm
F
H

(c)
J
26.18 cm
15.09 cm
K
L

(d) D
61°
F
17.91 cm
E

(e) Q   P   T
5.3 km
R   S

PRS is equilateral.
QTSR is a rectangle.
Write the angles on your copy of the diagram.

Fig. 24:11

(f) X
54.8°
Y
17.6 m
Z

How are you going to make a right-angled triangle in this isosceles triangle?

(g) P
17.08 cm
Q
12.65 cm
R

When a triangle is not right-angled, our trig. ratios cease to be true. However it is possible to obtain a rule, called the **sine rule**, that can be used when the triangle is not right-angled. You must, however, know one side and the angle opposite it, together with another side or angle.

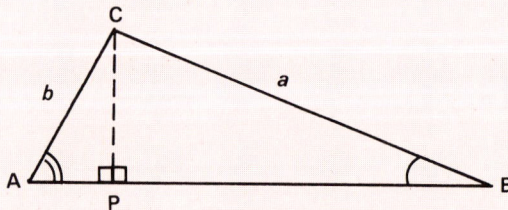

C
b
a
A
P
B

Fig. 24:12

In Figure 24:12, △ABC is not right-angled, but △APC and △CPB are. From these triangles:

$$CP = a \times \sin B \text{ and also } CP = b \times \sin A.$$

Therefore $a \times \sin B = b \times \sin A$. This is the **sine rule**, usually written:

$$\frac{a}{\sin A} = \frac{b}{\sin B} \quad \text{or} \quad \frac{\sin A}{a} = \frac{\sin B}{b}$$

to find a side   to find an angle

Always start the rule with the side, or the sine of the angle, that you are trying to find.

You can only use the sine rule if you know a side and the angle opposite to it.

Example In Figure 24:13,

$$\frac{x}{\sin 27°} = \frac{12\,\text{cm}}{\sin 48°}$$

$$x = \frac{12\,\text{cm}}{\sin 48°} \times \sin 27°$$

Key: 12 ÷ 48 sin × 27 sin =

Answer: $x = 7.33$ cm correct to 3 s.f.

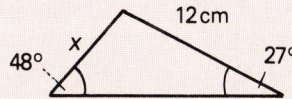

Fig. 24:13

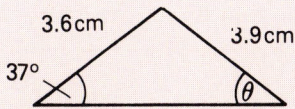

Fig. 24:14

Example In Figure 24:14,

$$\frac{\sin \theta}{3.6\,\text{cm}} = \frac{\sin 37°}{3.9\,\text{cm}}$$

$$\sin \theta = \frac{\sin 37°}{3.9\,\text{cm}} \times 3.6\,\text{cm}$$

Key: 37 sin ÷ 3.9 × 3.6 = ARCSIN

Answer: $\theta = 33.7°$ to the nearest 0.1°.

8 (a) In Figure 24:15 find:

    (i) x (ii) θ (iii) y.

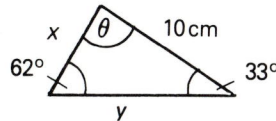

Fig. 24:15

(b) In Figure 24:16 find:

    (i) θ (ii) α (iii) x.

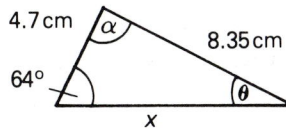

Fig. 24:16

(c) In Figure 24:17 find:

    (i) x (ii) θ (iii) y.

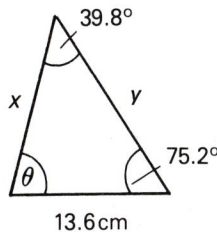

Fig. 24:17

9 Remind yourself of the division of degrees into minutes, explained on page 56.

Find all the sides and angles in:
(a) △ABC where ∠B = 40°9′, BC = 18.5 m, and AC = 12.6 m
(b) △DEF where ∠E = 46°18′, ∠F = 63°13′, and EF = 11.4 cm
(c) △GHI where ∠G = 78°41′, GI = 4.32 cm, and HI = 9.05 cm.

---

10 Using Pythagoras' Theorem find:
    (a) $\sin \theta$ if $\cos \theta = \frac{3}{5}$ (b) $\tan \theta$ if $\sin \theta = \frac{5}{13}$.

11 The sine of an obtuse angle, $\theta°$, is the same as the sine of $(180 - \theta)°$. For example, sin 150° = sin 30° = 0.5. Check this on your calculator, then find the acute angle whose sine is the same as:
(a) sin 100°   (b) sin 116.8°   (c) sin 152°12′.

12 In △ABC, ∠A = 48.3°, ∠C = 27.6°, and AC = 15.6 cm. Calculate:
    (a) ∠B   (b) BC   (c) AB.

13 Draw a graph of $y = \sin \theta$ and a graph of $y = \cos \theta$ using θ as the horizontal axis with θ at 10° intervals from 0° to 90°. You may be able to draw this curve using a computer, and your science department can probably produce one on an oscilloscope.

25 Factorisation: $x^2 \pm bx + c$

For Discussion

You have learnt how to expand two brackets to give a quadratic expression (Chapter 15).

$$(x + 4)(x + 3) \rightarrow x^2 + 3x + 4x + 12 \rightarrow x^2 + 7x + 12$$

By reversing this we can factorise a quadratic expression.

If **both signs** in the expression are + then the brackets will be (+)(+).

Example   To factorise $x^2 + 5x + 6$.

Work out the possible numbers to give the 6. They are 1×6 and 2×3. Try them:
$(x + 1)(x + 6) \rightarrow x^2 + 7x + 6$.   Wrong!
$(x + 2)(x + 3) \rightarrow x^2 + 5x + 6$.   Correct!

If the **middle sign** is − and the **last sign** is + then both brackets will have a minus sign: (−)(−).

Example   To factorise $x^2 - 5x + 6$.

$(x - 1)(x - 6) \rightarrow x^2 - 7x + 6$.   Wrong!
$(x - 2)(x - 3) \rightarrow x^2 - 5x + 6$.   Correct!

Factorise the expressions in questions 1 to 6.

1 (a) $x^2 + 2x + 1$ (b) $x^2 + 3x + 2$ (c) $x^2 + 7x + 6$
(d) $x^2 + 4x + 3$ (e) $x^2 + 8x + 12$ (f) $x^2 + 9x + 20$

2 (a) $x^2 - 2x + 1$ (b) $x^2 - 4x + 4$ (c) $x^2 - 3x + 2$
(d) $x^2 - 6x + 8$ (e) $x^2 - 6x + 9$ (f) $x^2 - 11x + 10$

3 (a) $x^2 + 4x + 4$ (b) $x^2 - 16x + 15$ (c) $x^2 - 5x + 4$
(d) $x^2 + 5x + 4$ (e) $x^2 + 7x + 12$ (f) $x^2 - 7x + 12$

4 (a) $x^2 + 20x + 51$ (b) $x^2 + 15x + 56$ (c) $x^2 - 16x + 48$
(d) $x^2 - 10x + 21$ (e) $x^2 + 11x + 10$ (f) $x^2 - 12x + 36$

5 (a) $x^2 + 14xy + 48y^2$ (b) $x^2 - 10xy + 24y^2$ (c) $x^2 + 10xy + 21y^2$
(d) $x^2 - 10xy + 16y^2$ (e) $x^2 - 4xy + 4y^2$ (f) $x^2 + 21xy + 68y^2$

6 (a) $a^2 - 9b^2$ (b) $p^2q^2 - 16$ (c) $x^4y^4 - 36$
(d) $bcx - bcy + mx - my$ (e) $4ax + 6bx + 6ay + 9by$

26 Area and volume: review

Areas of plane figures

Rectangle/Parallelogram   Base times height: $A = bh$

Triangle   Half base times height: $A = \frac{1}{2}bh$

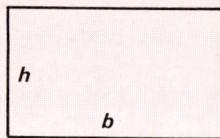 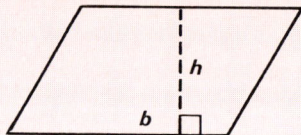 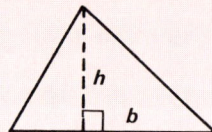

Rectangle   Parallelogram   Triangle

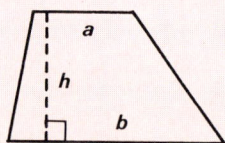 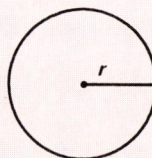

Fig. 26:1   Trapezium

Circle

Trapezium   Half the sum of the parallel sides times the distance between them: $A = \frac{1}{2}(a + b)h$

Circle   *Pi* times the square of the radius: $A = \pi r^2$

Surface areas of solids

Cylinder   Curved surface area is circumference of base circle times height of cylinder: πdh

                Total surface area is curved surface area plus the areas of the top and bottom circles: $\pi dh + 2\pi r^2$

Cone   Curved surface area is *pi* times the radius of the base times the slant height: πrl

                Total surface area is curved surface area plus the area of the base circle: $\pi rl + \pi r^2$

Sphere   4 times *pi* times the square of the radius: $4\pi r^2$

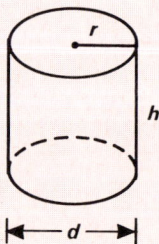  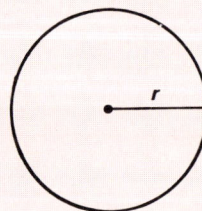

Fig. 26:2   Cylinder   Cone   Sphere

Volumes

Prism	A solid with a constant cross-section; that is, it has the same shape all through it. Examples: breeze block, kitchen roll, Toblerone box, unsharpened pencil, wedge of cheese.	Area of cross-section times length (or height): $V = Al$ or Ah
Cuboid	A prism with rectangular cross-sections, like a brick.	Length times width times height: $V = lwh$
Cylinder	A prism whose constant cross-section is a circle.	Area of circular cross-section times height: $V = \pi r^2 h$

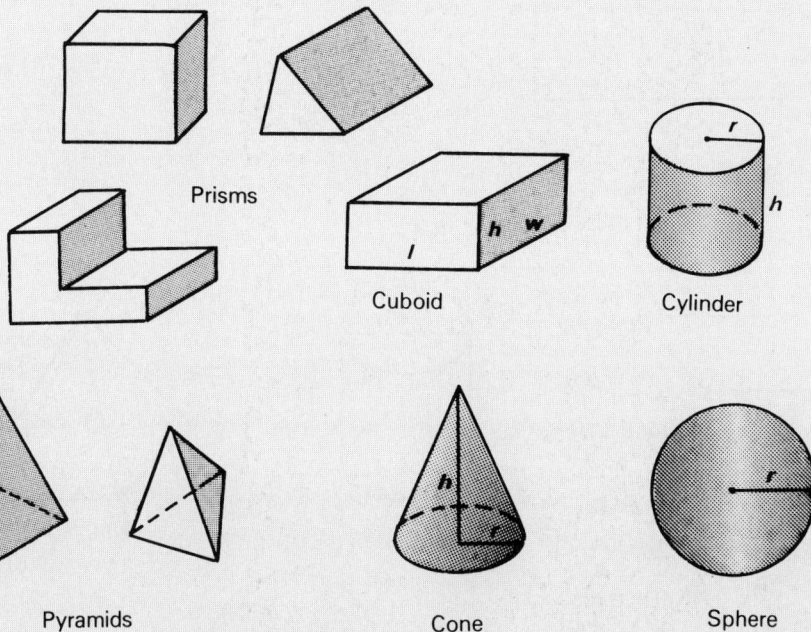

Prisms

Cuboid

Cylinder

Pyramids

Cone

Sphere

Fig. 26:3

Pyramid	May have any shape base, with sloping sides coming to a point. A pyramid on a square base is usually called an Egyptian pyramid. A pyramid with four triangular faces is called a tetrahedron (if the triangles are equilateral it is a regular tetrahedron).	One third of the base area times height: $V = \frac{1}{3}Ah$
Cone	A pyramid with a circular base.	One third of the circular base area times height: $V = \frac{1}{3}\pi r^2 h$
Sphere	A ball.	Four-thirds times pi times the cube of the radius: $V = \frac{4}{3}\pi r^3$

In all this exercise you should give answers correct to 3 significant figures, where appropriate.

1 Find the areas of the plane shapes in Figure 26:4.

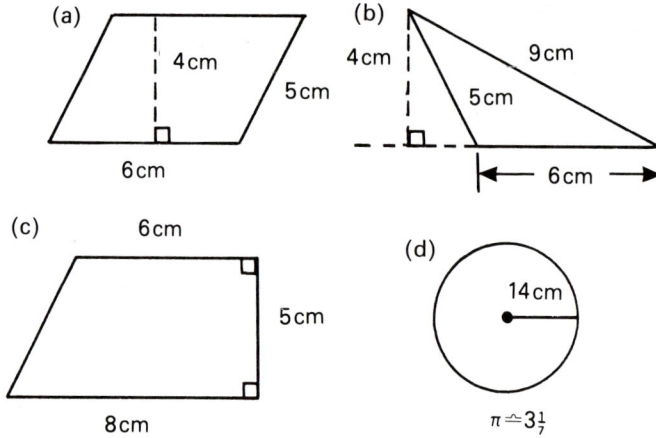

(a) 4 cm, 5 cm, 6 cm

(b) 4 cm, 9 cm, 5 cm, 6 cm

(c) 6 cm, 5 cm, 8 cm

(d) 14 cm, $\pi \simeq 3\frac{1}{7}$

Fig. 26:4

2 Find the volumes of the solids in Figure 26:5.

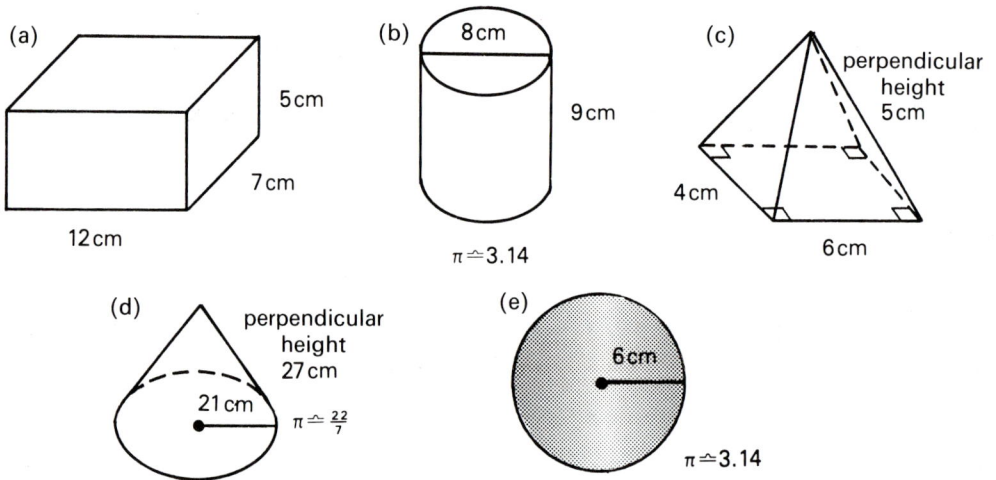

(a) 5 cm, 7 cm, 12 cm

(b) 8 cm, 9 cm, $\pi \simeq 3.14$

(c) perpendicular height 5 cm, 4 cm, 6 cm

(d) perpendicular height 27 cm, 21 cm, $\pi \simeq \frac{22}{7}$

(e) 6 cm, $\pi \simeq 3.14$

Fig. 26:5

*3 Find the areas of the plane shapes in Figure 26:6.

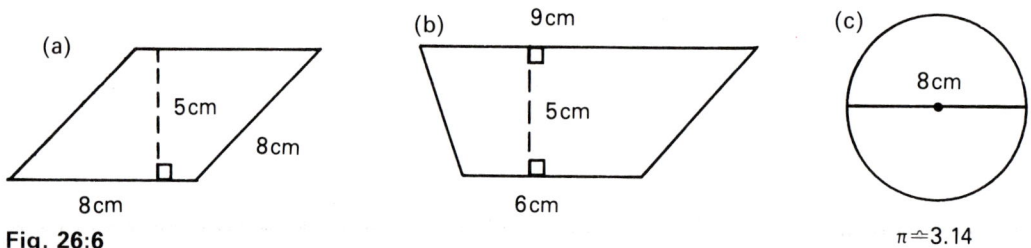

(a) 5 cm, 8 cm, 8 cm

(b) 9 cm, 5 cm, 6 cm

(c) 8 cm, $\pi \simeq 3.14$

Fig. 26:6

*4 Find the volumes of the solids in Figure 26:7.

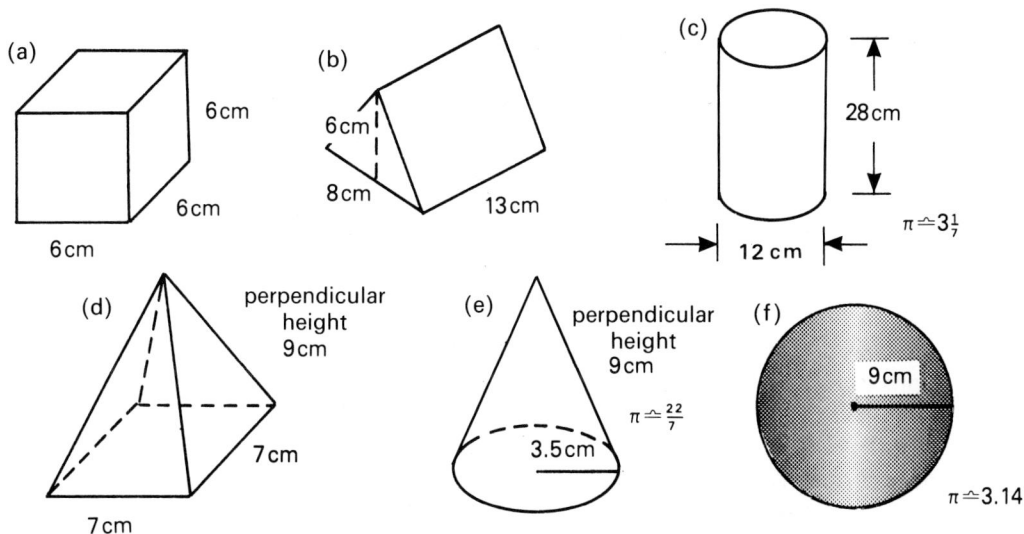

(a) 6cm, 6cm, 6cm

(b) 6cm, 8cm, 13cm

(c) 28cm, 12 cm, $\pi \simeq 3\frac{1}{7}$

(d) perpendicular height 9cm, 7cm, 7cm

(e) perpendicular height 9cm, 3.5cm, $\pi \simeq \frac{22}{7}$

(f) 9cm, $\pi \simeq 3.14$

Fig. 26:7

*5 A cylindrical barrel of radius 35 cm and height 120 cm is filled with cider. Taking $\pi = \frac{22}{7}$ find the volume of cider in the barrel in (a) cm³ and (b) litres. (Note: 1 litre = 1000 cm³.)

6 Figure 26:8 shows the plans and elevations of four solids. Calculate the volume of each solid in cm³.

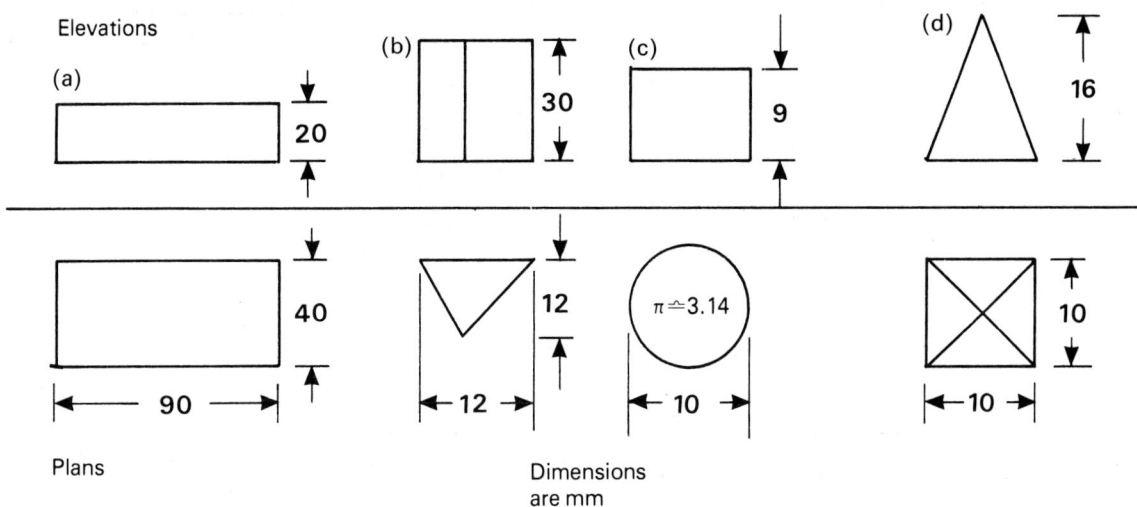

Elevations

(a) 20

(b) 30

(c) 9

(d) 16

40, 90

12, 12

$\pi \simeq 3.14$, 10

10, 10

Plans

Dimensions are mm

Fig. 26:8

7 An oil company drilled down 7000 m using a drill of 5 cm radius. How many m³ of earth and rock were removed? (Take $\pi = \frac{22}{7}$.)

155

8 The total surface area of a solid cylinder is made up of the curved surface area plus two circles.
$$A = \pi dh + 2\pi r^2 \to A = 2\pi rh + 2\pi r^2 \to 2\pi r(r + h)$$
The last version of this formula is the easiest one to use.

Show that the total surface area of a solid cone is $\pi r(r + l)$.

In questions 9 and 10 use a calculator *pi* key or take $\pi = 3.142$

9 Find the total surface area and the volume of a cylinder of radius 12 cm and height 23 cm.

10 Find the total surface area and the volume of a cone which fits exactly into the cylinder of question 9.

11 How many cans of radius 8 cm and height 30 cm can be filled from a drum of radius 40 cm and height 1.2 m? (Do not substitute a value for *pi*; cancel it out instead.)

12 A cylinder of metal, 20 cm long and of 2 cm radius, is drawn out into wire of 2 mm diameter. How long is the wire? (Do not substitute for *pi*.)

13 A cylindrical tank of diameter one metre and height 126 centimetres leaks through a rust hole at the rate of 5 litres per hour. How long will it take to lose half the contents of a full tank?

---

14 Water flows at 50 m/min through a pipe of radius 5 cm.
  (a) Find the cross-section area of the pipe.
  (b) Find how many cm³ of water flow through the pipe each minute.
  (c) Find the rate of flow in m³/h.

15 Find the rate of flow, in litres per hour correct to 3 sig. figs., in the following cases:
  (a) pipe of radius 7 cm; flow speed 32 m/min
  (b) pipe of radius 2 cm; flow speed 24 cm/s
  (c) pipe with rectangular cross-section, 8 cm × 6 cm; flow speed 102 m/min.

16 How many packets of butter 10 cm by 7 cm by 4 cm could be packed in a box 96 cm by 84 cm by 45 cm?

17 Write a computer program to calculate the areas and volumes of various geometric shapes.

---

27 Statistics: cumulative frequency curve

For Discussion

Marks	Frequency
1 to 5	1
6 to 10	2
11 to 15	4
16 to 20	7
21 to 25	9
26 to 30	5
31 to 35	2
36 to 40	1

Mark	Cumulative frequency
5 or less	1
10 or less	1 + 2 = 3
15 or less	3 + 4 = 7
20 or less	7 + 7 = 14
25 or less	14 + 9 = 23
30 or less	23 + 5 = 28
35 or less	28 + 2 = 30
40 or less	30 + 1 = 31

Marks out of 40 for a class of 31 pupils

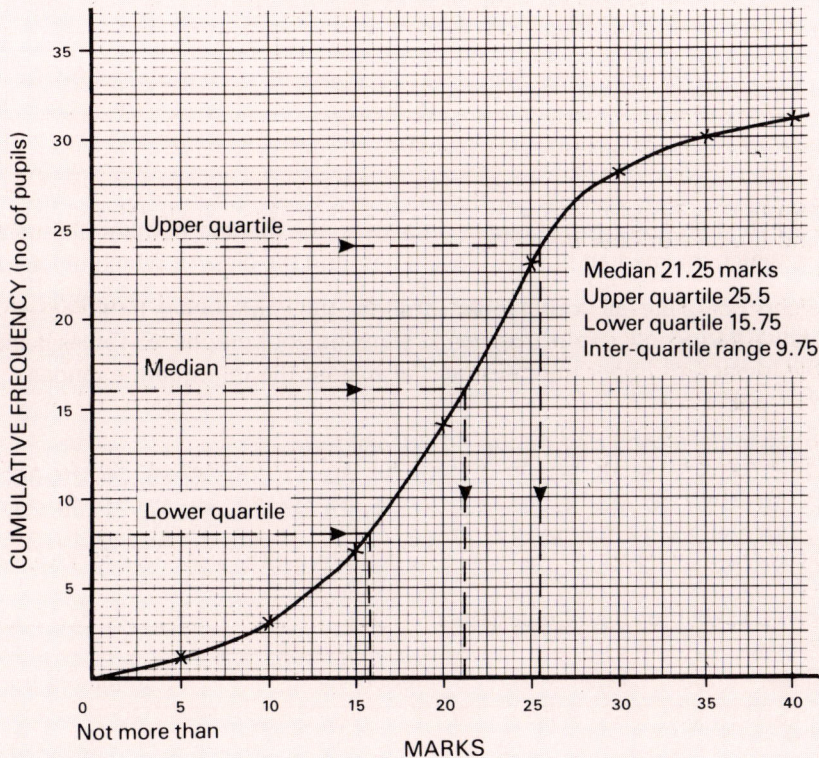

Median 21.25 marks
Upper quartile 25.5
Lower quartile 15.75
Inter-quartile range 9.75

Fig. 27:1

(a) What would you know about the test if the median had been: (i) 10   (ii) 30?

(b) What would you know about the class if the interquartiles had been at: (i) 5 and 35 (ii) 30 and 35?

1 **Table 1**

Flowers on a stem	0	1	2	3	4	5	6	7	8
Frequency	4	5	5	7	10	9	6	4	1
Cumulative frequency	4	9	14						

Table 2

Letters per word	1	2	3	4	5	6	7	8	9
Frequency	2	4	20	16	9	4	7	2	1
Cumulative frequency	2	6							

Table 3

Number absent	0–9	10–19	20–29	30–39	40–49	50–59
Frequency	6	11	24	16	9	3

Table 4

Weight (nearest kg)	40–45	46–50	51–55	56–60	61–65	66–70	71–75
Frequency	5	11	19	26	18	13	8

(a) State the median number of flowers on a stem, the median number of letters in a word, the class in which the median number absent lies, and the class in which the median weight lies.

(b) Copy and complete all four tables to show the cumulative frequencies, then draw the cumulative frequency graphs ('ogives'). Show on your graphs for Tables 3 and 4 the medians and the lower and upper quartiles, and also state the interquartile ranges.

*2 Number of absentees each day during an 80-day term:
16, 29, 41, 6, 24, 31, 19, 46, 30, 27, 22, 35, 34, 36, 11, 25, 38, 45, 26, 39, 36, 52, 12, 51, 24, 35, 47, 53, 28, 4, 37, 32, 44, 31, 15, 43, 20, 31, 29, 39, 22, 38, 42, 21, 38, 37, 34, 18, 33, 36, 34, 16, 42, 30, 28, 30, 37, 32, 20, 37, 27, 56, 43, 17, 57, 46, 33, 54, 47, 17, 39, 41, 48, 28, 35, 13, 31, 33, 21, 32.

(a) Complete a table for the above data:

Class	Tally	Frequency	Cumulative frequency
0–5
6–10
etc.			

(b) Draw a cumulative frequency curve to illustrate the data.

(c) Estimate from your curve the median number of days absent.

(d) Estimate the upper and lower quartiles and hence the interquartile range.

3 In Figure 27:2, line (a) shows that 3 pupils scored 10 marks or less and 28 scored more than 10. Write the information given by lines (b), (c) and (d).

Marks out of 40 for a class of 31 pupils

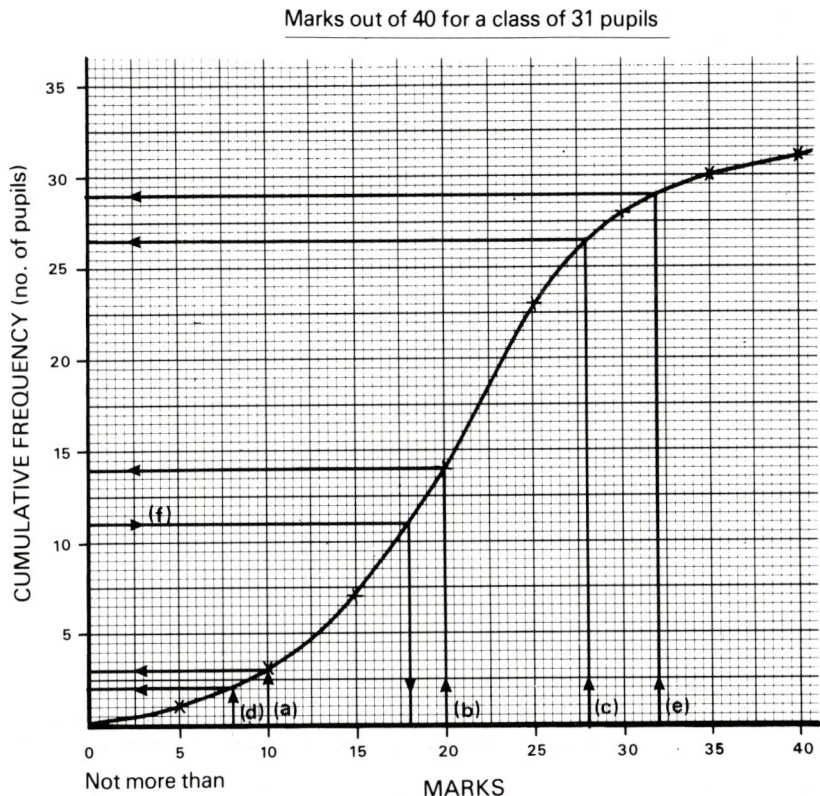

Fig. 27:2

4 **Example** Using Figure 27:2, estimate how many pupils scored more than 80%.

80% of 40 marks is 32 marks.
Line (e) at 32 marks shows that 29 pupils scored 80% or less, so 2 pupils scored more than 80%.

Note: Be careful not to use the number on the vertical axis as your answer; line (e) meets this axis at 29, so 31 − 29 = 2 pupils gives the correct answer.

(a) Estimate how many pupils scored:
   (i) 50% or less   (ii) more than 50%   (iii) 70% or less   (iv) more than 70%.

(b) Estimate the score of the pupil who came:
   (i) 10th   (ii) 20th   (iii) 25th.

5 Using Figure 27:2, state an estimated pass mark if the percentage of the pupils who are to pass is:
(a) 80%   (b) 60%.

27

6 The following tables show the percentages of the population of the United Kingdom in various age ranges in 1901 and 1981.

Age range	under 11	11–20	21–30	31–40	41–50	51–60	61–70	71–80	81–110
% of 1901 population	22	20	18	14	11	7	5	2	1
% of 1981 population	15	16	14	13	11	11	10	7	3

(a) Draw the cumulative frequency curve for 1901, plotting points above 11, 21, 31, etc., then find:
(i) the median age   (ii) the lower quartile   (ii) the upper quartile
(v) the interquartile range.

(b) Repeat part (a) for 1981, on the same axes.

(c) Comment on the differences between your two curves and two sets of answers. What differences would there be in the two societies?

7 Calculate the means for the data in question 1.

8 Find some test or exam results for different classes, and use statistical methods to compare the data. Write a report of your findings.

Reminders

A **vector** is a line with both length and direction. It is described by a column matrix, showing how far the end point of the vector is from the start point, measured horizontally and vertically. Positive and negative directions are the same as for graph axes. Figures 28:1 and 28:2 show two examples.

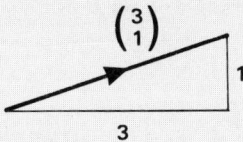

Fig. 28:1

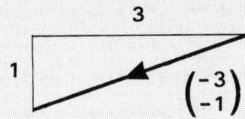

Fig. 28:2

Position vectors always start at the origin, so the position vector $\begin{pmatrix} -4 \\ 2 \end{pmatrix}$ starts at (0, 0) and ends at (−4, 2).

Shift vectors describe translations (slidings). In Figure 28:3 the hatched square is translated to the shaded square by the vector $\begin{pmatrix} 3 \\ 1 \end{pmatrix}$. Each corner of the square moves 3 units to the right and 1 unit upwards.

Shift vectors can start at any point, and the same vector can be at several places on the same grid.

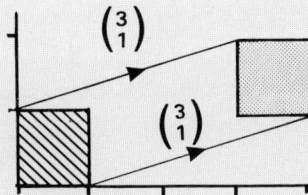

Fig. 28:3

Vectors may be described as a matrix, $\begin{pmatrix} 3 \\ 1 \end{pmatrix}$, or by their end letters, $\overrightarrow{AB}$, or by a single letter distinguished with a wavy line underneath (or sometimes, in print, by being thicker). The same letter (or letters) can be used for parallel vectors, but different letters *must* be used for non-parallel vectors.

Figures 28:4 to 28:7 show some examples of parallel vectors.

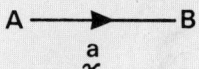

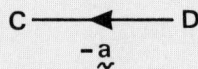

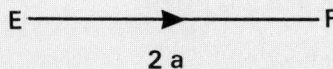

Fig. 28:4   Fig. 28:5   Fig. 28:6   Fig. 28:7

A **resultant vector** is the single vector that gives the same translation as all the others added together.

In Figure 28:8, $\overrightarrow{OA}$ is the resultant of $\underset{\sim}{a}$ and $\underset{\sim}{b}$ and $\underset{\sim}{c}$.

The resultant vector may be calculated by adding the matrices of the given set of vectors.

In Figure 28:8, $\overrightarrow{OA} = \underset{\sim}{a} + \underset{\sim}{b} + \underset{\sim}{c} = \begin{pmatrix} 4 \\ 0 \end{pmatrix} + \begin{pmatrix} -2 \\ -1 \end{pmatrix} + \begin{pmatrix} 0 \\ -3 \end{pmatrix} = \begin{pmatrix} 2 \\ -4 \end{pmatrix}.$

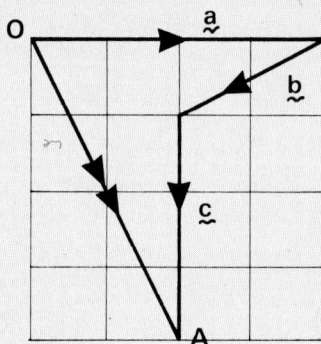

Fig. 28:8

The **magnitude of a vector** is its length. It is often written as $|\overrightarrow{OA}|$, and it is called the **modulus** of $\overrightarrow{OA}$.

In Figure 28:9

$(|\overrightarrow{OA}|)^2 = 2^2 + 4^2$ (Pythagoras' Theorem).

Hence $|\overrightarrow{OA}| = \sqrt{20} = 4.47$ to 3 s.f.

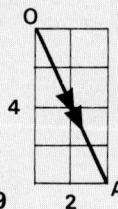

Fig. 28:9

1 If $\underset{\sim}{a} = \begin{pmatrix} 4 \\ 0 \end{pmatrix}$, $\underset{\sim}{b} = \begin{pmatrix} -2 \\ -1 \end{pmatrix}$, and $\underset{\sim}{c} = \begin{pmatrix} 0 \\ -3 \end{pmatrix}$, draw on squared paper the following vectors, starting each one from the end of the previous one. Remember the arrows!

$\underset{\sim}{a}$; $\underset{\sim}{b}$; $\underset{\sim}{c}$; $-\underset{\sim}{b}$; $\frac{1}{2}\underset{\sim}{a}$; $-\underset{\sim}{c}$; $2\underset{\sim}{a}$; $2\underset{\sim}{b}$; $-\frac{1}{2}\underset{\sim}{a}$; $\frac{1}{3}\underset{\sim}{c}$; $1\frac{1}{2}\underset{\sim}{a}$; $-2\underset{\sim}{b}$; $\underset{\sim}{c}$; $-4\underset{\sim}{a}$

*2 $\underset{\sim}{d} = \begin{pmatrix} 2 \\ 0 \end{pmatrix}$; $\underset{\sim}{e} = \begin{pmatrix} 0 \\ -2 \end{pmatrix}$; $\underset{\sim}{f} = \begin{pmatrix} 1 \\ 2 \end{pmatrix}$; $\underset{\sim}{g} = \begin{pmatrix} 3 \\ 2 \end{pmatrix}$; $\underset{\sim}{h} = \begin{pmatrix} -2 \\ 1 \end{pmatrix}$; $\underset{\sim}{i} = \begin{pmatrix} 1 \\ -2 \end{pmatrix}$

The resultant of $\underset{\sim}{d}$ and $\underset{\sim}{h}$ is $\begin{pmatrix} 2 \\ 0 \end{pmatrix} + \begin{pmatrix} -2 \\ 1 \end{pmatrix} = \begin{pmatrix} 0 \\ 1 \end{pmatrix}.$

Calculate the resultant of:
(a) $\underset{\sim}{d}$ and $\underset{\sim}{e}$ (b) $\underset{\sim}{e}$ and $\underset{\sim}{f}$ (c) $\underset{\sim}{e}$ and $\underset{\sim}{g}$ (d) $\underset{\sim}{e}$ and $\underset{\sim}{h}$ (e) $\underset{\sim}{d}$ and $\underset{\sim}{i}$
(f) $\underset{\sim}{d}$ and $\underset{\sim}{g}$ (g) $\underset{\sim}{f}$ and $\underset{\sim}{d}$ (h) $\underset{\sim}{i}$ and $\underset{\sim}{h}$ (i) $\underset{\sim}{g}$ and $\underset{\sim}{h}$ (j) $\underset{\sim}{i}$ and $\underset{\sim}{f}$.

*3 Figure 28:10 shows the resultant of vectors $\underset{\sim}{d}$ and $\underset{\sim}{h}$ (defined in question 2).

Draw on squared paper each pair of vectors given in parts (a) to (j) of question 2, and show their resultant. Check that your drawn answers agree with the ones you calculated.

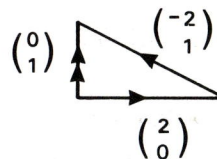

Fig. 28:10

*4 For the vectors given in question 2, calculate:
(a) $|\underset{\sim}{d}|$ (b) $|\underset{\sim}{e}|$ (c) $|\underset{\sim}{f}|$ (d) $|\underset{\sim}{g}|$ (e) $|\underset{\sim}{h}|$ (f) $|\underset{\sim}{i}|$.

5 (a) Show by calculation and drawing that the resultant of vectors $\underset{\sim}{e}$, $2\underset{\sim}{h}$, and $2\underset{\sim}{f}$ (as defined in question 2) is $-2\underset{\sim}{i}$.

(b) Vector $\overrightarrow{AB}$ is described by the matrix $\begin{pmatrix} 3 \\ -5 \end{pmatrix}$. Sketch this vector, then calculate $|\overrightarrow{AB}|$ correct to 3 s.f.

6 In Figure 28:11, state in terms of $\underset{\sim}{a}$ and $\underset{\sim}{b}$:
(a) $\overrightarrow{QP}$ (b) $\overrightarrow{SP}$ (c) $\overrightarrow{SR}$ (d) $\overrightarrow{RS}$
(e) $\overrightarrow{QR}$ (f) $\overrightarrow{RQ}$
(g) $\overrightarrow{PR}$ (the resultant of $\overrightarrow{PQ}$ and $\overrightarrow{PS}$).

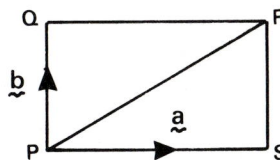

Fig. 28:11

7 In Figure 28:12, $\overrightarrow{SQ} = \overrightarrow{SP} + \overrightarrow{PQ}$, and $\overrightarrow{QS} = \overrightarrow{QP} + \overrightarrow{PS}$.
Write $\overrightarrow{SQ}$ and $\overrightarrow{QS}$ in terms of $\underset{\sim}{a}$ and $\underset{\sim}{b}$.

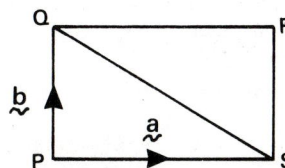

Fig. 28:12

8 In Figure 28:13, M is the midpoint of AC.
$\overrightarrow{BA} = \underset{\sim}{x}$, $\overrightarrow{BC} = \underset{\sim}{y}$, and $\overrightarrow{AM} = \underset{\sim}{z}$.

Express in terms of $\underset{\sim}{x}$, $\underset{\sim}{y}$ and $\underset{\sim}{z}$:
(a) $\overrightarrow{AB}$ (b) $\overrightarrow{CB}$ (c) $\overrightarrow{AC}$ (d) $\overrightarrow{MA}$ (e) $\overrightarrow{CM}$
(f) $\overrightarrow{BM}$ (in 2 ways)   (g) $\overrightarrow{MB}$ (in 2 ways).

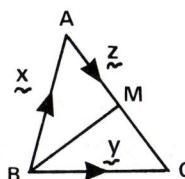

Fig. 28:13

9 In Figure 28:14, PQRS is a parallelogram. M is the midpoint of PS.

Express in terms of $\underset{\sim}{a}$ and $\underset{\sim}{b}$:
(a) $\overrightarrow{SP}$ (b) $\overrightarrow{SR}$ (c) $\overrightarrow{PM}$
(d) $\overrightarrow{QM}$ (e) $\overrightarrow{MR}$.

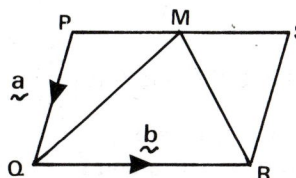

Fig. 28:14

10 KLMN is a square. D, E, F and G are the midpoints of KL, LM, MN and NK respectively. $\overrightarrow{GK} = \underset{\sim}{c}$ and $\overrightarrow{KD} = \underset{\sim}{d}$.

Express in terms of $\underset{\sim}{c}$ and $\underset{\sim}{d}$:

(a) $\overrightarrow{ML}$ (b) $\overrightarrow{LD}$ (c) $\overrightarrow{NF}$ (d) $\overrightarrow{GE}$ (e) $\overrightarrow{GD}$ (f) $\overrightarrow{GL}$ (g) $\overrightarrow{GM}$ (g) $\overrightarrow{GF}$.

11 State the missing vector:

(a) $\overrightarrow{AB} + \overrightarrow{BC} = \ldots$ (b) $\overrightarrow{AB} + \ldots = \overrightarrow{AF}$.

Illustrate your answers with diagrams.

12 $\overrightarrow{AB}$ and $\overrightarrow{AC}$ are at an angle of 30°, and AB = 2AC. Sketch $\overrightarrow{AB}$ and $\overrightarrow{AC}$ and their resultant vector.

13 If $\overrightarrow{AB} = \overrightarrow{CD}$ then AB and CD are equal and parallel. Sketch $\overrightarrow{AB}$ and $\overrightarrow{BC}$ if $\overrightarrow{AB} = \overrightarrow{BC}$.

14 ABC is a straight line with AB : AC = 1 : 3. If $\overrightarrow{AB} = \underset{\sim}{x}$, state in terms of $\underset{\sim}{x}$:

(a) $\overrightarrow{AC}$ (b) $\overrightarrow{BC}$.

15 (a) $\underset{\sim}{c}$ and $\underset{\sim}{d}$ are two sides of a triangle. Explain why $\underset{\sim}{c} \neq \underset{\sim}{d}$.

(b) If in part (a), $(h + k)\underset{\sim}{c} = (h - k + 1)\underset{\sim}{d}$, calculate h and k.

16 $\overrightarrow{AB}$ and $\overrightarrow{AC}$ are two vectors. M is the midpoint of BC. Express $\overrightarrow{BC}$ and $\overrightarrow{BM}$ in terms of $\overrightarrow{AB}$ and $\overrightarrow{AC}$, and hence show that $\overrightarrow{AM} = \frac{1}{2}(\overrightarrow{AB} + \overrightarrow{AC})$.

This is an important theorem about two vectors.

---

17 (a) If $\overrightarrow{XY} = h\overrightarrow{BC}$, what is true about $\overrightarrow{XY}$ and $\overrightarrow{BC}$?

(b) X and Y are points on the sides AB and AC of △ABC such that $\dfrac{AX}{AB} = \dfrac{AY}{AC} = h$. It

follows that AX = hAB and that $\overrightarrow{AX} = h\overrightarrow{AB}$.

Show that $\overrightarrow{XY} = h\overrightarrow{BC}$ and hence that XY // BC and $\dfrac{XY}{BC} = h$.

18 Prove that when the midpoints of the sides of a parallelogram are joined in order they form another parallelogram.

As $(x - 2)(2x + 1) \rightarrow 2x^2 - 3x - 2$, it follows that $2x^2 - 3x - 2 \rightarrow (x - 2)(2x + 1)$.

By no means all quadratic expressions factorise into two brackets, but an examiner will not ask you to factorise one that will not.

There is no 'golden rule' which will give you the correct answer first time, but the following will help:

(a) If the last sign is + then both brackets have the sign of the middle term. (In checking with 'FOIL', O + I gives the middle term.)

    Examples $x^2 + 10x + 21 \rightarrow (x + 3)(x + 7)$
                 Check: $+7x + 3x \rightarrow 10x$
                       $+3 \times +7 \rightarrow +21$

                  $x^2 - 10x + 21 \rightarrow (x - 3)(x - 7)$
                 Check: $-7x - 3x \rightarrow -10x$
                       $-3 \times -7 \rightarrow +21$

(b) If the last sign is − then one bracket is + and the other is −, and you must be very careful to get the correct sign with each number, for example $(2x + 1)(x - 2)$ gives $2x^2 - 3x - 2$, whilst $(2x - 1)(x + 2)$ gives $2x^2 + 3x - 2$.

    Examples $x^2 + 4x - 12 \rightarrow (x + 6)(x - 2)$
                 Check: $-2x + 6x \rightarrow +4x$
                       $+6 \times -2 \rightarrow -12$

                  $x^2 - 4x - 12 \rightarrow (x - 6)(x + 2)$
                 Check: $+2x - 6x \rightarrow -4x$
                       $-6 \times +2 \rightarrow -12$

(c) If there are a lot of factors to choose from, start with the pair closest together and work upwards, e.g. for 12 try 3×4, then 2×6, then 1×12.

Factorise:

1   (a) $x^2 + x - 2$ (b) $x^2 - 3x - 4$ (c) $x^2 + x - 6$ (d) $x^2 - 5x - 6$
    (e) $x^2 + x - 12$

2   (a) $x^2 - 4x - 12$ (b) $x^2 - 8x - 20$ (c) $x^2 + 3x - 18$ (d) $x^2 - 7x - 18$
    (e) $x^2 - 5x - 36$

3   (a) $x^2 + 13x + 36$ (b) $x^2 + 9x - 36$ (c) $x^2 + 5x - 36$ (d) $x^2 + 16x - 36$
    (e) $x^2 - 15x + 36$

4   **Example** $2x^2 - 5x - 3 \rightarrow (2x + 1)(x - 3)$
                 Check: $-6x + 1x \rightarrow -5x$
                      $+1 \times -3 \rightarrow -3$

    (a) $2x^2 + 3x + 1$ (b) $2x^2 - 3x + 1$ (c) $2x^2 - x - 3$ (d) $2x^2 + x - 3$

*5   (a) $2x^2 - 9x + 4$ (b) $2x^2 - 7x + 3$ (c) $3x^2 - 2x - 16$ (d) $3x^2 + 2x - 8$

6 (a) $2x^2 - 5x + 2$ (b) $3x^2 + 7x + 2$ (c) $2x^2 - x - 1$ (d) $x^2 + 2x - 8$

7 (a) $3x^2 + 2x - 16$ (b) $2x^2 - 3x - 35$ (c) $3x^2 - 7x - 6$ (d) $3x^2 - 2x - 16$

8 (a) $2x^2 - 21x - 36$ (b) $2x^2 + 7x - 30$ (c) $x^2 + 17x + 30$ (d) $2x^2 - x - 36$

9 (a) $2x^2 - 7x - 30$ (b) $2x^2 - 17x + 30$ (c) $3x^2 - 71x - 24$ (d) $2x^2 + 11x - 30$

Where the coefficient of the first term is not prime then there is a choice for the start of each bracket. Once again you just have to try possibilities until you hit on the right one. You will find that you get cleverer at deciding the most likely possibilities.

Example $6x^2 + 11x - 10$ could start $(3x \quad)(2x \quad)$ or $(x \quad)(6x \quad)$.
By trial we find that $6x^2 + 11x - 10 \rightarrow (3x - 2)(2x + 5)$.

If the constant term (the number on the end) is prime it is better to put this in first, then work out the start of the bracket.

Example   To factorise $12x^2 - 17x - 5$.

$12x^2$ could come from $3x \times 4x$, or $2x \times 6x$, or $1x \times 12x$.
But 5 must come from 1×5.
So start with $(\quad 1)(\quad 5)$ and then try the possibilities to give $12x^2$.

The answer is $(3x - 5)(4x + 1)$.

10 (a) $4x^2 + 3x - 1$ (b) $4x^2 + 4x + 1$ (c) $6x^2 + 13x + 6$ (d) $6x^2 - 7x - 24$

11 (a) $6x^2 - 11x + 4$ (b) $6x^2 + 11x + 4$ (c) $6x^2 - 5x - 4$ (d) $6x^2 + 5x - 4$

12 Not all quadratic expressions will factorise. For example, $x^2 + 2x + 12$.

Factorise the following where possible:
(a) $x^2 - x - 6$ (b) $x^2 - 5x + 6$ (c) $x^2 - 3x + 6$ (d) $2x^2 + 7x - 35$
(e) $2x^2 + 7x - 6$ (f) $3x^2 - x - 2$ (g) $3x^2 + 11x + 6$ (h) $3x^2 - 3x + 8$
(i) $x^2 + 5x - 11$ (j) $3x^2 - 9x - 7$ (k) $4x^2 + 5x + 1$ (l) $6x^2 + 13x + 6$
(m) $3x^2 + 13x + 12$ (n) $6x^2 - 23x - 4$ (o) $8x^2 + 14x + 3$ (p) $8x^2 - 2x - 15$.

13 Always check for common factors before factorising by quadratic brackets.

Example $8x^2 + 4x - 12 \rightarrow 4(2x^2 + x - 3) \rightarrow 4(2x + 3)(x - 1)$

Factorise:
(a) $3x^2 + 12x + 12$ (b) $2x^2 + 14x + 24$ (c) $3x^2 - 15x + 12$
(d) $9x^2 + 6x - 3$ (e) $4x^2 - 10x + 4$ (f) $12x^2 - 8x - 64$.

14 Factorise:
(a) $9a^2 - 6ab + b^2$ (b) $35 - 12x + x^2$ (c) $35 - 2x - x^2$
(d) $3x^2 + 5xy - 2y^2$ (e) $8x^2 - 18x + 9$ (f) $25x^2 - 30xy + 9y^2$
(g) $x^2 + x + \frac{1}{4}$ (h) $x^2 + x - \frac{3}{4}$ (i) $24 - 2x - x^2$

15 Is it possible to program a computer to factorise?

A Paying interest

Investment Account

FAST LOANS

How is interest calculated?

ENDOWMENT (25 YEAR) POLICIES

HIGHER INTEREST ACCOUNT

RATES OF INTEREST

How is the interest paid?

Investments

YEARS AFTER PURCHASE	VALUE AT END OF YEAR	YIELD FOR YEAR TAX-FREE
1	£26.44	5.76%
2	£28.20	6.66%
3	£30.40	7.80%
4	£33.12	8.95%
5	£36.48	10.14%

High interest

7.00% NET = 10.00% GROSS EQUIVALENT

PAID-UP SHARES

MONTHLY SAVINGS PLAN

8.00% NET = 11.43% GROSS EQUIVALENT

- Access to capital at any time
- Competitive interest

home improvement loans

What is the interest rate?

The return is equivalent to a compound annual interest rate of 7.85% over the full five years.

Interest is paid to someone who lends money. If you lend money to the Government through your Post Office Savings Bank then the Government will pay you interest.

If the interest rate is $5\frac{1}{2}$% per annum (p.a.) then the Government will pay you £5.50 a year for every £100 you lend them.

If the interest is not added to the loan (that is, it is always 'withdrawn') it is called **simple interest**.

If the interest *is* added to the loan it is called **compound interest**. Most everyday-life interest is compound interest.

Examples   £100 loaned at 5% p.a. simple interest for 3 years gives £5 each year, making £15 interest altogether.

£100 loaned at 5% p.a. compound interest for 3 years gives £5 interest the first year, but 5% of £105 = £5.25 the second year, and 5% of £110.25 = £5.51 the third year, a total of £15.76 interest.

There are two formulae which you can use, but they are not essential:

$$\text{Simple interest} = \frac{P \times R \times T}{100} \qquad \text{Compound interest} = P\left(1 + \frac{R}{100}\right)^{T} - P$$

where P is the principal (the initial amount lent)
R is the interest rate p.a.
T is the number of years for which the principal is lent.

For Discussion

Lender	Borrower	Interest rate
Tom Smith	Government (P.O. Savings Bank)	5% p.a.
Building Society	Jill Brown (Mortgage)	13% p.a.
Alan Watson	Building Society (Share Account)	10% p.a.
Town Bank	Jack Tar (Bank Loan)	14% p.a.
Andrew White	Town Bank (Deposit Account)	11% p.a.
Joy Jones	City Council (Local Authority Bonds)	15% p.a.

(a) Why does Tom lend his money to the Government?

(b)   (i) Why does the Building Society lend money to Jill?
    (ii) What is meant by 'a 100% mortgage'?
    (iii) Where does the Building Society get the money that it lends to Jill?
    (iv) Why does the Building Society charge Jill 13% interest but only pay Alan 10%?

(c) The City Council will not accept loans of less than £1000, nor for periods of less than one year. All interest is taxable. The Post Office Savings Bank will accept any amount over 50p for any time over 7 days. No tax is payable if less than £1400 is loaned. In what way is this reflected in the interests paid?

Using a calculator
To find the interest on £50 at 8%.

Key: 50 $\boxed{\times}$ 8 $\boxed{\%}$
or   50 $\boxed{\times}$ 8 $\boxed{\%}$ $\boxed{=}$
or   50 $\boxed{\times}$ 0.08 $\boxed{=}$

1   As £1 = 100p, then 1% interest is 1p in the £, and 5% interest is 5p in the £.

If Tom puts £50 in his Post Office Savings Bank Account for 1 year, how much interest will he receive?

2   13% interest is £13 in every £100.

Jill borrows £10 000 from the Building Society. How much interest will she pay in a year?

3   Alan Watson lends £500 to the Building Society through his Share Account. How much interest will they pay him at the end of the year?

4   Jack Tar borrows £800 from Town Bank. How much interest will he have to pay them in a year?

5   Andrew White puts £75 into his Town Bank Deposit Account. How much interest will he receive after a year?

*6 Copy and complete the following table.

PRINCIPAL (the amount loaned)	P	£10	£20	£35	£400	£350
RATE (the rate of interest p.a.)	R	5%	5%	5%	11%	10%
INTEREST (the amount paid in a year)	I					

P	£1000	£750	£650	£100	£200	£10 000	£600	£1600
R	9%	5%	5%	$7\frac{1}{2}$%	$10\frac{1}{2}$%	$10\frac{1}{2}$%	$10\frac{1}{2}$%	10%
I								

7 Find (a) to (l) in the following table.

PRINCIPAL (the amount loaned)	P	£1500	£5000	£2000	£350	(e)
RATE (the rate of interest p.a.)	R	$8\frac{1}{2}$%	$13\frac{1}{4}$%	(c)	(d)	12%
INTEREST (the amount paid in a year)	I	(a)	(b)	£160	£24.50	£300

P	(f)	£400	(h)	(i)	£360	£150	(l)
R	$7\frac{1}{2}$%	(g)	11%	$10\frac{1}{2}$%	$9\frac{3}{4}$%	(k)	13%
I	£112.50	£32	£2.75	£262.50	(j)	£5.25	£1625

8 How much will Fred need to put into his Building Society account to gain £750 in a year if the interest rate is 10%?

9 How much interest will Joy Jones actually receive if she lends £3000 at 15%, but has to pay tax at 30% on the interest?

10 Jim buys a £4000 car on Hire Purchase for £150 deposit, then 12 monthly payments of £360. June buys a similar car, borrowing the money for one year from a bank at 11% interest. She agrees to pay the £4000 plus interest back in 12 equal monthly instalments.

(a) How much does Jim pay altogether for his car?

(b) How much will June have to pay her bank each month?

(c) Who pays less for their car, and by how much?

11 Write a brief essay on one, or both, of the following.

(a) Building Societies.

(b) You win £100 000 on the pools. In what different ways might you invest it to give a regular income? What are the advantages and disadvantages of each possibility?

B Compound interest

When the interest due to lenders is calculated, it is not usually sent to them. Instead, it is added to the amount they have loaned. This means that each interest payment is larger than the previous one (if the principal is not reduced). We say that the interest is 'compounded', hence **compound interest**.

Example To find the result of investing £1000 for 3 years at 10% compound interest.

Date	Details	Interest	Principal
31 Jan 1985	1st principal	–	£1000
31 Jan 1986	1st interest (10%)	£100	£1100
31 Jan 1987	2nd interest	£110	£1210
31 Jan 1988	3rd interest	£121	£1331

Using a calculator

The $\boxed{M+}$ key can be used to gradually increase the principal. Either a % key may be used, or the fact that, e.g., 8% is 0.08.

Example To find the principal and interest paid each year on £100 saved at 8% compound interest.

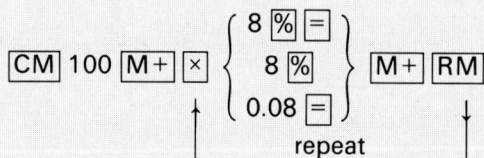

$$\boxed{CM}\ 100\ \boxed{M+}\ \boxed{\times} \left\{ \begin{array}{l} 8\ \boxed{\%}\ \boxed{=} \\ 8\ \boxed{\%} \\ 0.08\ \boxed{=} \end{array} \right\} \boxed{M+}\ \boxed{RM}$$

repeat

Alternatively, using the fact that to increase by r% you multiply by $(100 + r)$% and to decrease by r% you multiply by $(100 - r)$%, the following sequences may be used.

Compound interest at 8% on £100:

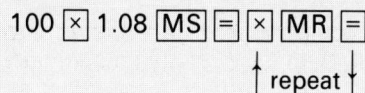

$$100\ \boxed{\times}\ 1.08\ \boxed{MS}\ \boxed{=}\ \boxed{\times}\ \boxed{MR}\ \boxed{=}$$

repeat

Depreciation (reduction in value) at 8% per annum on £10 000:

$$10\,000\ \boxed{\times}\ 0.92\ \boxed{MS}\ \boxed{=}\ \boxed{\times}\ \boxed{MR}\ \boxed{=}$$

repeat

1 Find how much the following principals will amount to in 3 years if the interest is compounded.

	(a)	(b)	(c)	(d)	(e)
Principal	£3000	£400	£200	£250	£1000
Rate	10%	5%	7%	6%	7%

*2 Find how much the following principals will amount to if the interest is compounded.

	(a)	(b)	(c)	(d)
Principal	£1000	£100	£900	£4000
Rate	6%	5%	10%	5%
Time	2 yrs	2 yrs	3 yrs	3 yrs

3 Using a calculator to give each year's interest correct to the nearest penny, find the total amount for the following:

	(a)	(b)	(c)	(d)
Principal	£500	£80	£25	£10 000
Rate	7%	12%	$7\frac{1}{2}$%	$10\frac{1}{2}$%
Time	4 yrs	4 yrs	3 yrs	4 yrs

4 Bigtown Building Society pays savers 5% interest, correct to the nearest penny, every six months, this interest being compounded. If I put £500 into the Building Society for 3 years, calculate each of the six interest payments.

5 Ali reckons that the value of his new car will depreciate (go down) each year by about 5% of its previous year's value. If his car cost £10 000 when new, what would Ali expect it to be worth at the end of each of the next five years?

---

6 The formula $A = P(1 + 0.01R)^T$ gives the amount (A) which a principal (P) will realise in T years at R% per year compound interest.

Use a calculator to find, correct to the nearest 1p, the amount A for the following.

	(a)	(b)	(c)	(d)	(e)
P pounds	500	750	600	1000	30 000
R% p.a.	7%	10%	$12\frac{1}{2}$%	11%	13%
T years	5	8	10	4	6

7 (a) Jill wishes to buy a house and the City Building Society agree to lend her £25 000 at 15% compound interest for 25 years. If Jill made no payments until the 25 years were up, how much would she then owe the Society?

(b) The answer to part (a) should show you that this would be an impossible way to borrow the money. Instead, the Society would expect Jill either to pay all the interest due each year, and cover the £25 000 debt with life insurance (this is called an endowment mortgage); or to pay the interest and refund some of the loan each year, so that after 25 years she would owe them nothing.
If the society asks Jill to pay £303 each month for 25 years to discharge the debt completely, how much would she pay altogether?

(c) In reality, the interest paid is reduced by an income tax allowance, so the endowment mortgage payment would be about £210 a month, plus a life insurance premium. How much would the house really have cost her? (Remember that the house will almost certainly be worth a lot more than £25 000 in 25 years' time.)

8 Using the formula given in question 6 to help you, design a computer program to advise customers on their investments and loans.

Using your calculator

Wildlife in peril

Poachers and the modern materialistic society are decimating the wildlife of Africa. People want ivory for ornaments, or for a hedge against inflation; rhino horn for dagger handles, or powdered for medicine; zebra skins for drums and mats; antelope antlers and ostrich eggs for souvenirs; giraffe tails for fly-whisks; cheetah pelts for coats; even gorilla hands for ash-trays.

In 1971, Uganda's Rwenzori National Park had 3000 elephants, 18 000 buffaloes, and 10 500 hippos. Ten years later Idi Amin's army had reduced the numbers to 160 elephants, 4000 buffaloes, and 2900 hippos.

In 1982, 40 000 elephants were killed in Africa, the ivory selling at £17 a pound (a tusk can weigh as much as 100 pounds). Britain alone imported £400 000-worth of tusks, as well as 111 863 items of carved ivory.

Between 1973 and 1983, black rhinos were reduced by 80% to about 10 000, Kenya's reduction being from 16 000 to 1000. North Yemen alone imported 20 tons of horn for traditional daggers, representing 8000 dead rhinos. A rhino's horn could bring an African peasant £300, perhaps thirty times his annual income; yet when powdered for medicine it would fetch £450 an ounce.

It is not only poaching that threatens the wildlife. Parkland is in increasing demand for farming; in 1983 Kenya's population of 17 million was growing at 4% per annum.

Unless every African country establishes heavily policed reserves, like Nairobi's 44 sq.mile National Park, where 10 000 animals live in safety, it is only a matter of time before many African animals will only be found in zoos.

Use the above information and your calculator to answer the following questions.

1   What is meant by 'a hedge against inflation'?

2   What does 'decimating' mean?

3   Percentage reduction = $\dfrac{\text{reduction}}{\text{original number}} \times 100\%$.
    What was the percentage reduction in Rwenzori Park between 1971 and 1981 of:
    (a) elephants   (b) buffaloes   (c) hippos?

4   About how much was paid for the ivory from the elephants killed in 1982? (Give the answer in millions of pounds.)

5   About how many tusks did Britain import in 1982?

6   About how many rhinos lived in Africa in 1973?

7   What was the percentage reduction in Kenya's rhino population between 1973 and 1983?

8   What is the average weight of one rhino's horn in pounds? (2240 pounds = 1 ton)

(*Zoological Society of London*)

9 Percentage profit = $\dfrac{\text{profit}}{\text{original cost}} \times 100\%$.

What is the percentage profit on a ton of powdered rhino horn?
(16 ounces = 1 pound; 2240 pounds = 1 ton)

10 If Kenya's population continued to grow at the 1983 rate, what would it have been in 1986 (after 3 years of increase)?

11 What is the area of Nairobi's National Park in:
(a) square kilometres   (b) hectares?
(Take 5 miles = 8 kilometres. One hectare = $10\,000\,\text{m}^2$.)

12 How many square metres does each animal in the National Park have, on average?

A   Solution by factorisation

For Discussion

1   Find x if:
   (a) $3x = 0$ (b) $xy = 0$ when $y = 5$ (c) $xy = 0$ when $y \neq 0$.

2   What can you say about x and y if $xy = 0$?

3   Find x if:
   (a) $4x = 0$ (b) $x - 4 = 0$ (c) $x + 8 = 0$ (d) $3x + 15 = 0$
   (e) $2x - 3 = 0$ (f) $6x + 3 = 0$ (g) $4x + 12 = 0$ (h) $2x - 5 = 0$
   (i) $3x - 2 = 0$.

If two factors multiply to give zero, one or both of them must be zero.

Example   If $x(x - 4) = 0$ then either $x = 0$ or $x - 4 = 0$.
   Hence $x(x - 4) = 0$ when $x = 0$ and when $x = 4$.

Example   If $(x - 2)(x + 4) = 0$ then either $x - 2 = 0$ or $x + 4 = 0$.
   Hence $(x - 2)(x + 4) = 0$ when $x = 2$ and when $x = -4$.

Example   If $(2x - 7)(4x + 2) = 0$ then either $2x - 7 = 0$ or $4x + 2 = 0$.
   If $2x - 7 = 0$ then $2x = 7$, so $x = 3\frac{1}{2}$ is one solution.
   If $4x + 2 = 0$ then $4x = -2$, so $x = -\frac{1}{2}$ is the other solution.

Find all possible solutions to the equations in questions 1 to 10.

1   (a) $(x - 2)(x - 3) = 0$ (b) $(x + 6)(x - 2) = 0$ (c) $(x - 4)(x + 5) = 0$

2   (a) $(x - 2)^2 = 0$ (b) $(x + 4)^2 = 0$ (c) $(2x - 6)(x - 5) = 0$

3   (a) $(x - 3)(2x - 3) = 0$ (b) $(x + 4)(2x + 3) = 0$ (c) $(4x - 2)(x + 3) = 0$

4   (a) $(2x + 9)(x - 7) = 0$ (b) $(3x - 1)(x + 2) = 0$ (c) $(2x - 3)^2 = 0$

5   **Example**   Solve $x^2 - 3x + 2 = 0$.

   $x^2 - 3x + 2 = 0 \rightarrow (x - 2)(x - 1) = 0$
   $\therefore x = 2$ and $x = 1$ are the solutions.

   Example   Solve $x^2 - 4x = 0$.

   Quadratics with no constant term are easy to factorise by taking out the common factor. Make a special point of remembering this.

   $x^2 - 4x = 0 \rightarrow x(x - 4) = 0$
   $\therefore x = 0$ and $x = 4$ are solutions.

Example Find the roots of $2x^2 + 3x - 9 = 0$.

The **roots** of an equation are the values of x that make it true.

$2x^2 + 3x - 9 = 0 \rightarrow (2x - 3)(x + 3) = 0$
The roots are $x = 1\frac{1}{2}$ and $x = -3$.

Solve:
(a) $x^2 - 4x + 4 = 0$ (b) $x^2 - 2x + 1 = 0$ (c) $a^2 - 6a + 9 = 0$

6 (a) $n^2 - 7n + 10 = 0$ (b) $2w^2 + w - 3 = 0$ (c) $2t^2 - t - 3 = 0$

7 (a) $2x^2 - 5x + 3 = 0$ (b) $a^2 - 4 = 0$ (c) $x^2 - 49 = 0$

8 (a) $f^2 + 7f = 0$ (b) $2e^2 - 5e = 0$ (c) $2d^2 + 7d = 0$

*9 (a) $x^2 - 5x + 4 = 0$ (b) $s^2 - 11s + 28 = 0$ (c) $h^2 - 8h + 15 = 0$

*10 (a) $2t^2 - 7t - 15 = 0$ (b) $2x^2 - 3x - 5 = 0$ (c) $2k^2 + k - 6 = 0$

11 Rearrange the following equation terms to make equations that equal zero, then find their solutions.
(a) $x^2 = 25$ (b) $h^2 - 4h = 32$ (c) $w^2 - w = 90$ (d) $x^2 - 22 = -9x$
(e) $m^2 + 6 = 7m$ (f) $7x^2 = 3x$

12 Example The length of a rectangle is 5 cm more than its breadth. If its area is 24 cm² find its dimensions.

Let its breadth be x cm.
Then its length is $x + 5$ cm.
Area of rectangle = length × breadth
so $x(x + 5) = 24 \rightarrow x^2 + 5x = 24$
$\rightarrow x^2 + 5x - 24 = 0 \rightarrow (x - 3)(x + 8) = 0$
$\therefore x = 3$ or $x = -8$.

24 cm² x cm

$(x + 5)$ cm

A dimension cannot be -8, so the only solution is $x = 3$, making the breadth 3 cm and the length 8 cm.

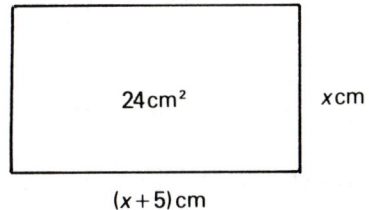

(a) A room is 8 metres longer than it is wide, and its area is 20 m². Use a similar method to the example to find its dimensions.

(b) A square of paper is reduced by 5 cm in one direction and by 1 cm in the other direction, leaving a rectangle of area 60 cm². How long was a side of the original square? (Start by letting this be x cm.)

(c) The square of a number added to ten times itself equals 39. What could the number be?

(d) Ten times the square of a fraction is nine more than forty-three times the fraction. What could the fraction be? (Start by letting it be f.)

13 Find the equation whose solutions are:
(a) 2 and 4   (b) -2 and 2   (c) -3 and 5   (d) -2 and -4
(e) -4 and 0   (f) $\frac{1}{2}$ and $-\frac{2}{3}$.

14 State the co-ordinates of the crossing points on the x- and y-axes for the parabolic graphs whose equations are given.

Remember that $y = ax^2 + bx + c$ must cross the y-axis at $y = c$.

(a) $y = x^2 - 2x - 3$ (b) $y = x^2 - 3x - 4$ (c) $y = x^2 - 9$
(d) $y = x^2 + 4x + 4$ (e) $y = 2x^2 - 5x + 3$ (f) $y = x^2 + 6x + 9$

15 Use your answers to question 14 to sketch the parabolas.

16 (a) Find the side of a square whose area doubles when its length is increased by 6 cm and its breadth is increased by 4 cm.

(b) The ages of two children are 4 years and 8 years. In how many years' time will the product of their ages be 165?

(c) The sum of the first n natural numbers is $\frac{1}{2}n(n + 1)$.
What is n if the sum is 210?

B Harder equations

Many quadratic equations will not factorise; the following two methods may then be used. Your teacher will explain them fully.

Method One Completing the Square

Example Solve $x^2 + 5x - 1 = 0$.

$$x^2 + 5x - 1 = 0 \rightarrow \left(x + \frac{5}{2}\right)^2 - \frac{25}{4} - 1 = 0$$

$$\left[\text{Note: } \left(x + \frac{5}{2}\right)\left(x + \frac{5}{2}\right) = x^2 + 5x + \frac{25}{4}\right]$$

$$\rightarrow \left(x + \frac{5}{2}\right)^2 - \frac{29}{4} = 0$$

$$\rightarrow \left(x + \frac{5}{2}\right)^2 = \frac{29}{4} = 7.25$$

$$\rightarrow x + 2.5 = \sqrt{7.25}$$

$$\rightarrow x + 2.5 \simeq \pm 2.69 \text{ [plus or minus 2.69]}$$

So $x \simeq -2.5 + 2.69 = \underline{0.19}$
or $x \simeq -2.5 - 2.69 = \underline{-5.19}$

Example Solve $3x^2 - 2x - 2 = 0$.

$$3x^2 - 2x - 2 = 0 \xrightarrow{\div \text{ all by 3}} x^2 - \tfrac{2}{3}x - \tfrac{2}{3} = 0 \qquad \text{Note: We divide by 3 to reduce}$$
the coefficient of x^2 to unity.

$$\rightarrow (x - \tfrac{1}{3})^2 - \tfrac{1}{9} - \tfrac{2}{3} = 0$$

$$\rightarrow (x - \tfrac{1}{3})^2 - \tfrac{7}{9} = 0$$

$$\rightarrow (x - \tfrac{1}{3})^2 = \tfrac{7}{9} = 0.\dot{7}$$

$$\rightarrow x - 0.\dot{3} = \sqrt{0.\dot{7}}$$

$$\rightarrow x - 0.\dot{3} \simeq \pm 0.882$$

Either $x \simeq 0.\dot{3} + 0.882 \simeq \underline{1.22}$

or $x \simeq 0.\dot{3} - 0.882 \simeq \underline{-0.55}$

Method Two The Quadratic Equation Formula

The formula is obtained by applying the completion of the square method to the general quadratic equation $ax^2 + bx + c = 0$. Your teacher will illustrate this.

$$x = \frac{-b \pm \sqrt{b^2 - 4ac}}{2a}$$

Learn: x equals minus b plus or minus the square root of b squared minus four ac, **all** divided by two a.

Example Solve $x^2 + 5x - 1 = 0$.

$a = 1 \qquad b = 5 \qquad c = -1$

$$x = \frac{-5 \pm \sqrt{25 + 4}}{2} = \frac{-5 \pm \sqrt{29}}{2}$$

$$x \simeq \frac{-5 \pm 5.385}{2}$$

Solutions are $x \simeq 0.19$ and $x \simeq -5.19$

Example Solve $3x^2 - 2x - 2 = 0$.

$a = 3 \qquad b = -2 \qquad c = -2$

$$x = \frac{+2 \pm \sqrt{4 + 24}}{6} = \frac{2 \pm \sqrt{28}}{6}$$

$$x \simeq \frac{2 \pm 5.29}{6}$$

Solutions are $x \simeq 1.22$ and $x \simeq -0.55$

Solve the following equations, either by completing the square, or by using the formula, or both. Your teacher will tell you which to do.

1 (a) $x^2 + 4x - 1 = 0$ (b) $x^2 - 2x - 5 = 0$ (c) $x^2 + 5x + 3 = 0$

2 (a) $x^2 - 7x - 12 = 0$ (b) $x^2 + 8x + 9 = 0$ (c) $x^2 + x - 3 = 0$

3 (a) $2x^2 + 2x - 1 = 0$ (b) $2x^2 - 4x + 1 = 0$ (c) $3x^2 + 5x - 3 = 0$

*4 (a) $x^2 - 3x - 3 = 0$ (b) $2x^2 + 4x + 1 = 0$ (c) $3x^2 + x - 1 = 0$

5 Rearrange the following equations to make them equal zero. Do not solve the equations.
   (a) $2x^2 - 4x = 1$ (b) $3x^2 = 4x - 1$ (c) $2x^2 - 2 = x$ (d) $3 - 4x = 2x^2$
   (e) $3x - 3 = x^2$ (f) $x^2 + 4 = 4x$

6 Solve the following equations.
   (a) $2x(x + 4) = 1$ (b) $x^2 = 5(1 - x)$ (c) $3(x^2 - 1) = 4x$

---

7 **Example** Solve $\dfrac{x^2}{2} - x - \dfrac{3}{4} = 0$

First remove the fractions by multiplying every term by 4, which is the common denominator.

As $4 \times \dfrac{x^2}{2} \to 2x^2$, and $4 \times \dfrac{3}{4} \to 3$, then the equation becomes:

$2x^2 - 4x - 3 = 0$.

This can now be solved as before.

Write each of the following as an equation without fractions. Do not solve the equations.

   (a) $\dfrac{x^2}{3} + x + \dfrac{7}{9} = 0$ (b) $2x^2 - \dfrac{3x}{4} + \dfrac{1}{2} = 0$ (c) $\dfrac{3x^2 - 2x}{5} + 1 = 0$

   (d) $x^2 - \dfrac{2x + 1}{3} = 0$ (e) $\dfrac{x(3x - 2)}{4} + \dfrac{1}{3} = \dfrac{1}{4}$ (f) $\dfrac{(2x + 1)(x - 1)}{2} = \dfrac{3x}{4}$

8 $ax^2 + bx + c = 0$ is solved by $x = \dfrac{-b \pm \sqrt{(b^2 - 4ac)}}{2a}$.

   (a) Investigate what will happen if $b^2 < 4ac$. Illustrate your answer with a graph of $y = x^2 + x + 1$, being used to try to solve the equation $x^2 + x + 1 = 0$.

   (b) Investigate what will happen if $b^2 = 4ac$. Illustrate your answer with a graph of $y = x^2 + 2x + 1$, being used to solve $x^2 + 2x + 1 = 0$.

   (c) State which of the equations in the answers to questions 5 and 7 have:
      A solutions   B no solutions   C equal roots

9 The solutions of the equation $ax^2 + bx + c = 0$ are called its 'roots'. These roots are usually referred to as α (alpha) and β (beta).

$$\alpha + \beta = -\frac{b}{a} \quad \text{and} \quad \alpha\beta = \frac{c}{a}$$

These two facts can be used to check solutions to a quadratic equation. Use them to check your answers to questions 2 and 3.

10 Write a computer program which will solve a quadratic equation, or inform you that there are no solutions.

A Solid symmetry

For Discussion

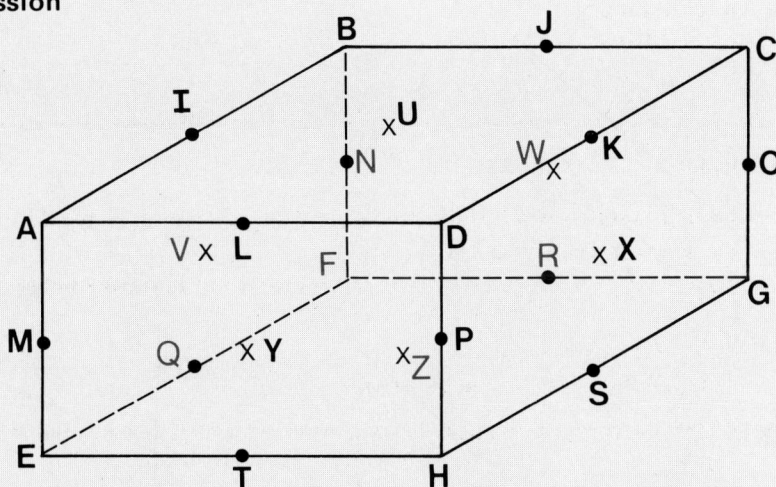

Fig. 32:1

Figure 32:1 shows a cuboid. ABCD is a square.
● are the midpoints of edges; X are the centres of faces.
Red lettered points are seen 'through' the cuboid.

Planes of symmetry (Reflection symmetry)
In the cuboid shown in Figure 32:1, BDHF is a plane of symmetry (see Figure 32:2), and IKSQ is another (see Figure 32:3).

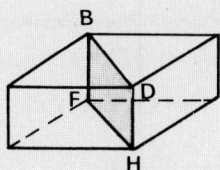

Fig. 32:2

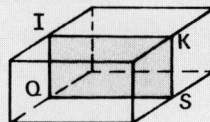

Fig. 32:3

A plane of symmetry acts as a mirror.

Why is ADGF in Figure 32:1 not a plane of symmetry?

State the 2 diagonal planes of symmetry, and the 3 planes of symmetry parallel to the faces, for the cuboid in Figure 32:1.

Axes of symmetry (Rotational symmetry)
In Figure 32:1, UZ is an axis of symmetry (see Figure 32:4). The cuboid in Figure 32:1 has 3 axes of symmetry. Name each of them.

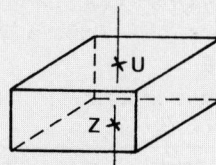

Fig. 32:4

Symmetry number
The symmetry number of a solid is the number of ways that it can be put into a mould of itself. The cuboid in Figure 32:1 has symmetry number 8.

1 For the solids in Figure 32:5 state the planes of symmetry and the axes of symmetry. Also state the symmetry numbers.

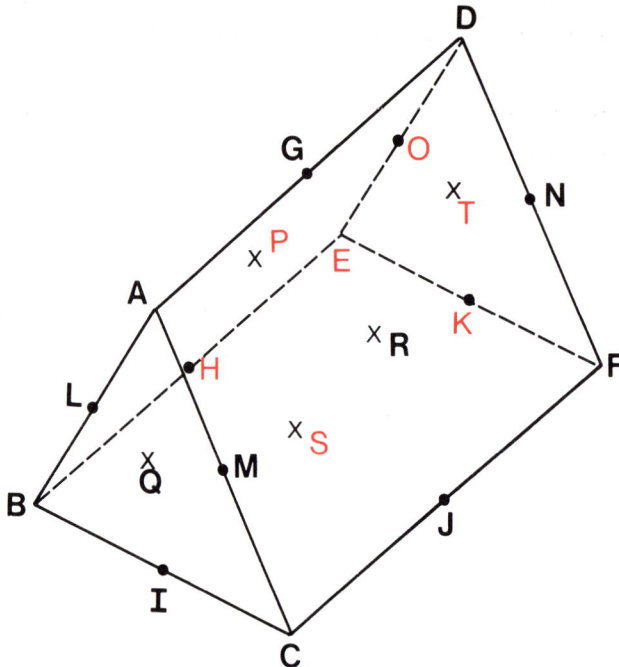

(a) Equilateral triangular prism:
   4 planes of symmetry
   4 axes of symmetry

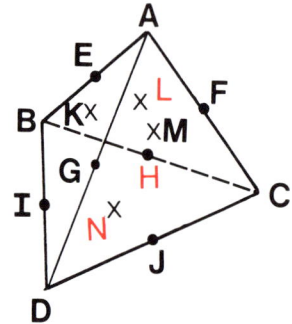

(b) Tetrahedron with 4
   triangular faces:
   6 planes of symmetry
   7 axes of symmetry

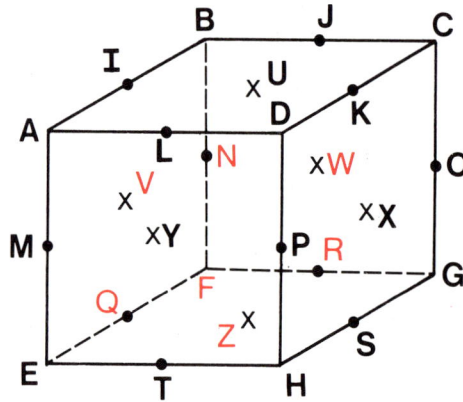

(c) Cube:
   9 planes of symmetry
   13 axes of symmetry

Fig. 32:5

2 Investigate the symmetry of:
   (a) a regular pentagonal prism
   (b) a regular hexagonal prism
   (c) a frustum of a square-based pyramid.

B Mensuration of solids

When answering questions about solids, spend some time trying to see the two-dimensional object drawn on the paper as a real three-dimensional solid. You may even be able to make a simple model of it from a piece of paper.

Calculations are always made using plane shapes (often triangles) and it is sensible to draw these triangles separately (and fairly accurately).

1 Figure 32:6 shows a square-based pyramid. All edges are 6 cm long.

   (a) How many (i) faces, (ii) vertices (corners), and (iii) edges, has the solid?

   (b) In Figure 32:6 there are 9 right angles, although none of them looks like a right angle. Find them all, e.g. ∠VMA.

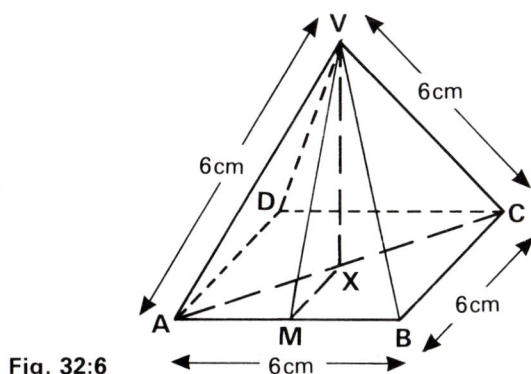

Fig. 32:6

   (c) Sketch the following triangles, showing the right angles as true ones, and marking any known lengths. The first is shown as an example in Figure 32:7.
   (i) ΔABC   (ii) ΔADC   (iii) ΔVAM
   (iv) ΔVMB   (v) ΔVAX   (vi) ΔVXM

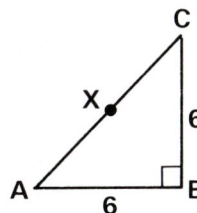

Fig. 32:7 True shape of ΔABC

   (d) Use Pythagoras' Theorem to help you calculate, correct to 3 significant figures:
   (i) AC (use ΔABC)   (ii) AX   (iii) XM   (iv) VM (use ΔVAM)
   (v) VX (use ΔVXM).

2 Referring to Figure 32:6, and using the triangles drawn for question 1(c), calculate to the nearest 0.1°:
   (a) ∠BAC   (b) ∠CAD   (c) ∠VAM   (d) ∠MVA   (e) ∠VBM   (f) ∠BVM
   (g) ∠VAX   (h) ∠AVX   (i) ∠VMX   (j) ∠XVM.

*3 Repeat question 1(d) for a square-based pyramid with all edges 7 cm long.

*4 Repeat question 2 for the 7 cm-sided pyramid.

5 Calculate the volume of the pyramid in Figure 32:6.

6 Calculate the volume of the triangular prism in Figure 32:10 (see question 13).

7 In Figure 32:6, what fraction of the volume of the pyramid is VAMX?

8 Calculate the volume of water in a pool 30 metres long and 10 metres wide, which is 4 metres deep at the deep end and 1 metre at the shallow end, the bottom sloping along the whole 30 metre length. (Give your answer in litres in standard form.)

9 (a) Figure 32:8 shows a cuboid. Copy the diagram carefully.

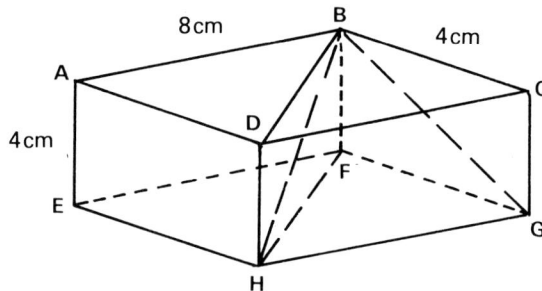

Fig. 32:8

(b) Each of the following triangles is right-angled. Draw a sketch of each triangle, clearly showing its right angle as a true one.
  (i) ΔBCD   (ii) ΔHDB   (iii) ΔBHF   (iv) ΔBGH.

(c) If BC = 4 cm, CD = 8 cm, and DH = 4 cm, use triangles (i) and (ii) in part (b) to calculate:
  (i) BD   (ii) BH (use BD^2 from part (a))   (iii) ∠BDC   (iv) ∠DBH.

10 (a) Copy Figure 32:9, then draw separate sketches of triangles EHG and AEG, showing clearly their right angles as true ones.

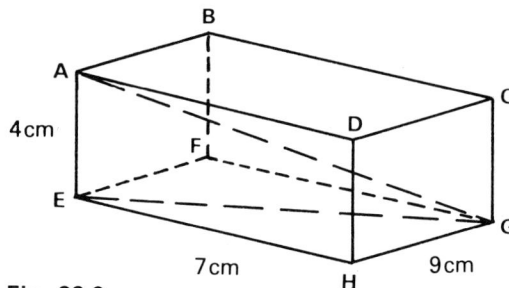

Fig. 32:9

(b) Calculate AG if EH = 7 cm, GH = 9 cm, and AE = 4 cm. (Hint: Find EG^2 but not EG.)

(c) Calculate ∠AGE.

11 Calculate the longest diagonal of a 4 cm cube.

12 Can a 3.7 metre rod be put in a box 3 m long, 2 m wide, and 1 m deep?

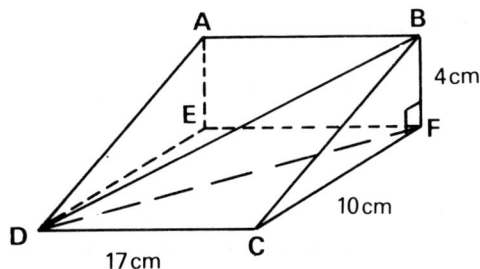

Fig. 32:10

13 (a) Copy the diagram of a wedge (Figure 32:10).

   (b) Sketch △BCF, right-angled at F, and calculate BC.

   (c) Sketch △FCD, right-angled at C, and calculate FD.

   (d) Sketch △BFD, right-angled at F, and calculate BD. (Hint: Use FD^2 but do not find FD.)

   (e) Calculate ∠BCF and ∠BDF. These are called the angles of greatest slope and of least slope.

---

14 On your copy of Figure 32:10 mark a point M, the midpoint of DC. Which of the following angles is the biggest and which is the smallest (or are they all the same size!)?
   (a) ∠BDF   (b) ∠BMF   (c) ∠BCF

15 In question 14, ∠BCF is called 'the angle between the planes ABCD and EFCD'. Note that both BC and CF are at right angles to DC, the line of intersection of the two planes.

   ABCD is the tailboard of a removal van, with AB = 2 metres and BC = 1 metre. AB is hinged to the van at a point 60 cm above the ground, and CD lies on the ground.

   Calculate:
   (a) the angle of greatest slope of the tailboard
   (b) the angle of least slope of the tailboard.

16 Calculate the side of the largest cube that could be cut from a 10 cm diameter sphere.

17 If the pyramid in question 1 is cut by a plane parallel to, and 2 cm from, its base, calculate the volume of the lower part (a 'frustum').

Papers

Paper One

1   A cube is painted on all faces, then cut into 27 identical smaller cubes. How many of the smaller cubes are painted on:
    (a) three faces   (b) two faces   (c) one face   (d) no face?

2   Julia travels 9 miles in 45 minutes. How far would she travel in 3 hours at the same average speed?

3   Joy spends three-quarters of her money and then has 15p left. How much did she have to start with?

4   (a) 3 h 16 min + 48 min   (b) 6 h − 3 h 27 min

5   What should be the next two numbers in the following sequences?
    (a) 3, 5, 8, 12, 17   (b) 5, 9, 17, 33, 65   (c) 0, 3, 8, 15, 24

6   Figure P1 is a travel graph showing a journey made by a cycling club.

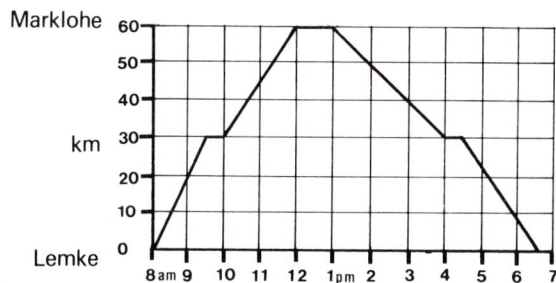

Fig. P1

    (a) When did they take the first break?
    (b) How long was the break at Marklohe?
    (c) What was the average speed from Lemke to Marklohe?
    (d) How long did they spend resting altogether?
    (e) How far did they cycle altogether?

*7   Sketch a compass showing the eight cardinal points N, E, S, W, NE, SE, SW and NW.

*8   How many degrees, turning clockwise, between:
    (a) E and W   (b) SE and W   (c) NW and SE   (d) NE and SE   (e) E and NE?

9   Between 1243 and 1423 a train covers 80 miles. What is its average speed?

10 Find two consecutive numbers such that one-third of the smaller is greater by 3 than one-quarter of the larger.

Paper Two

1 State the co-ordinates of the crossing point of:
(a) $y = 2$ and $x = 6$ (b) $y = 2$ and $y = x$.

2 On one pair of axes, from -4 to 4 each, draw the lines with equations $y = 3$, $y = -x$, and $y = 2x + 1$.

3 Multiply:
(a) $(6 \quad 0) \begin{pmatrix} 3 & 2 \\ 1 & 5 \end{pmatrix}$ (b) $\begin{pmatrix} 2 & 1 \\ 3 & 4 \end{pmatrix} \begin{pmatrix} 2 \\ 3 \end{pmatrix}$ (c) $\begin{pmatrix} 1 & 0 \\ 0 & 2 \end{pmatrix} \begin{pmatrix} 1 & -2 \\ 3 & 4 \end{pmatrix}$

4 Solve to find the number n:
(a) $3n = 12$ (b) $4n = -12$ (c) $3n - 5 = 4$ (d) $n + 8 = 4$.

5 If twelve tickets cost £2.52, how many could you buy for £4?

6 A lawn 25 m by 15 m is surrounded by a concrete path 2 m wide. What is the area of the path? (A diagram may help you.)

*7 Write in 12-hour-clock time:
(a) 1500   (b) 0842   (c) 1426.

*8 (a) 3 min 48 s + 51 s + 2 min 6 s   (b) 3 min 21 s − 1 min 51 s

9 Alan, Bill and Colin went fishing. Alan caught 5 more fish than Colin, and Bill caught 4 less than Alan and Colin together. If in all they caught 58 fish, how many did they each catch?

10 Solve simultaneously: $2x + y = 3$ and $x + 2y = 0$.

11 (a) Construct the grid shown in Figure P2.
(b) On your grid draw the set of cubes shown in Figure P3.
(c) How many cubes are needed to build Figure P3?

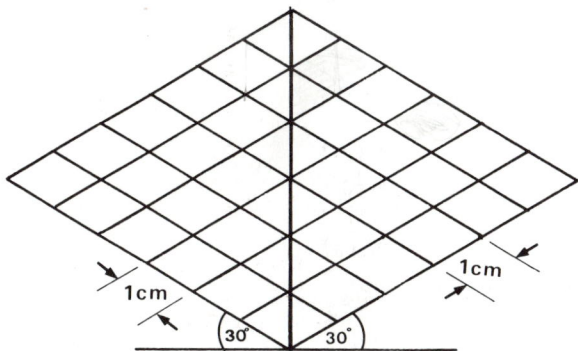

Fig. P2

Fig. P3

Paper Three

1 ξ = {letters in the word *prince*}. List a possible set A and a possible set B if n(A) = 2, n(B) = 2 and n(A $\cap$ B) = 1. Also state n(A′) and n(A $\cup$ B).

2 Of 55 pupils, 19 had been to Scotland and 31 to Wales. Twelve had not been to either.

Illustrate this information on a Venn diagram, then state how many had been to both countries.

3 Remove the brackets, then simplify:
   (a) 4(2a − 3) + 6a − 4   (b) −(2a + 4) + 3(2a − 5).

4 What is the equation of the straight line joining (0, 2) to (−5, 2)?

5 Draw four pairs of axes, each having scales from −2 to 4. Draw on each grid the quadrilateral given by the matrix $\begin{pmatrix} 0 & 2 & 2 & 1 \\ 0 & 0 & 1 & 1 \end{pmatrix}$.

Illustrate, answering one part on each grid, the effect on your shape of the transformation matrix:
   (a) $\begin{pmatrix} 2 & 0 \\ 0 & 2 \end{pmatrix}$ (b) $\begin{pmatrix} 0 & -1 \\ 1 & 0 \end{pmatrix}$ (c) $\begin{pmatrix} -1 & 0 \\ 0 & -1 \end{pmatrix}$ (d) $\begin{pmatrix} 0 & 1 \\ -1 & 0 \end{pmatrix}$

*6 Change to seconds:
   (a) 3 min 16 s   (b) 7 min 48 s.

7 Using the quadrilateral given in question 5, or otherwise, describe the transformation effected by:
   (a) $\begin{pmatrix} 1 & 2 \\ 0 & 1 \end{pmatrix}$ (b) $\begin{pmatrix} -1 & 0 \\ -1 & 1 \end{pmatrix}$ (c) $\begin{pmatrix} \frac{1}{2} & 0 \\ 0 & \frac{1}{2} \end{pmatrix}$

8 Find a number such that when 8 is added to it the result is twice as much as when 4 is subtracted from it.

9 Solve simultaneously, 5y − 4x = 1 and 3x − 10y = 18.

Paper Four

1 Use a calculator to find:
   (a) $\dfrac{628 \times 17}{6.8}$ (b) $\dfrac{648 \times 31.4}{726 \times 11.3}$ correct to 2 decimal places.

2 If $H = \dfrac{PLAN}{33\,000}$, calculate H when P = 66, L = 5, A = 5 and N = 40.

3 Remove the brackets then simplify:
   (a) 2(b − 3) + 4(b − 5)   (b) 3(b + 3) + 2(b − 4)   (c) −2(3 + b) + 2(b − 4).

4 In Figure P4
   ξ = {Alpha sports club}
   B = {badminton players}
   H = {hockey players}
   T = {tennis players}.

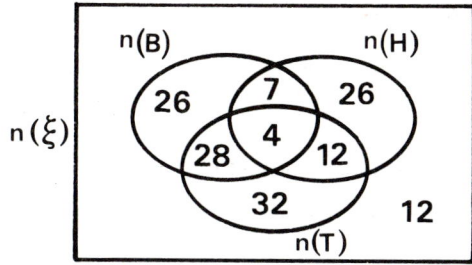

   Fig. P4

   State the number of people:
   (a) in the Alpha sports club
   (b) who play badminton
   (c) who only play hockey
   (d) who play two, but not three, of the available sports.

5 (a) 3 h 16 min + 48 min + 2 h 38 min   (b) 6 h 24 min − 3 h 37 min

*6 Solve simultaneously:
   (a) $x + y = 2$
        $3x - y = 14$ (Add)

   (b) $x + 3y = 11$ (Multiply top equation by 3,
        $3x + 2y = 12$ then subtract)

*7 Copy Figure P5, then shade the region A′.

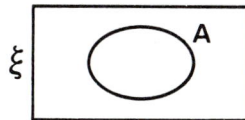

   Fig. P5

8 Solve to find h:
   (a) $3(2h - 4) = 3h - 6$ (b) $2(h + 3) = h + 9$.

9 Solve simultaneously: $2x - y = 8$ and $3x + 2y = 5$.

10 Using a calculator, find correct to 3 significant figures:
   (a) $\dfrac{40.28 \times 27.64}{3.64}$ (b) $\dfrac{39.374 \times 16.8^2}{4.9^3}$

11 Each of the n teams in a league are to play each of the others once. State, in terms of n, the number of matches required.

Paper Five

1 Using compasses, construct a triangle with sides 6 cm, 6 cm and 7 cm.

   Now use compasses to bisect each angle and so find the centre of the incircle. Draw this incircle, and measure its radius.

2 Draw two triangles which satisfy the conditions:
   AB = 6 cm, BC = 4 cm, $\angle$A = 35°. (Use a protractor for the angle.)

   In each triangle measure the size of angle C.

3 Discuss which design, Figure P6 or Figure P7, would produce a stronger gate.

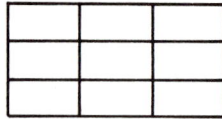

Fig. P6

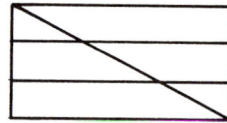

Fig. P7

4 Find the value of f if:
(a) $5(f - 3) = 15$ (b) $5(f - 3) + 2(2f - 5) = 11$.

5 How many chickens and how many cows are there in a farmyard if the total number of heads is 64 and the total number of feet is 158?

***6** Change to hours and minutes:
(a) 408 min   (b) 311 min.

7 Multiply:
(a) $\begin{pmatrix} 3 & 2 \\ 4 & 7 \end{pmatrix} \begin{pmatrix} 5 \\ 1 \end{pmatrix}$ (b) $\begin{pmatrix} -3 & 4 \\ 5 & -2 \end{pmatrix} \begin{pmatrix} -1 \\ 2 \end{pmatrix}$ (c) $(2 \quad 4) \begin{pmatrix} 3 & 4 \\ 1 & 2 \end{pmatrix}$
(d) $(-3 \quad 4) \begin{pmatrix} 2 & -4 \\ -3 & 3 \end{pmatrix}$

8 Twenty-six people answered a questionnaire about holiday travel. Twenty had travelled by air, 17 by ship, 18 by car, 16 by air and ship, 13 by ship and car, 15 by air and car, and 2 had not been on holiday. How many had:
(a) used all three means of transport
(b) been by car only
(c) been by air and car, but not by ship?

Paper Six

1 Write the following numbers approximated to 3 significant figures.
(a) 3042   (b) 42.38   (c) 1.234   (d) 3004   (e) 20.48   (f) 8.999

2 Write as an integer:
(a) 3.26×10^2
(b) the number displayed on a calculator as 3.06   03

3 Write in standard form ($A \times 10^x$ where $1 \leqslant A < 10$ and x is integral):
(a) 625   (b) 72 000   (c) 72 460.

4 Of thirty-eight pupils in a music group, 14 play wind instruments and 20 play string instruments. If 4 play both kinds of instruments, how many play neither?

5 In Figure P8, the two triangles are similar, as they have the same angles. Side AC corresponds (is in the same position relative to the angles) to side DE, so the sides of the two triangles are in the ratio 4 : 12.

(a) Write the ratio 4 : 12 in its simplest form.

(b) Calculate the length of: (i) FE   (ii) AB.

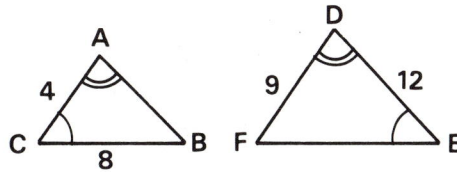

Fig. P8

(c) Calculate the unknown sides in Figure P9.

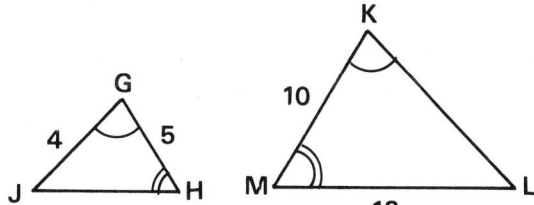

Fig. P9

*6 Write in 12-hour-clock time (including a.m. or p.m.):
(a) 0500   (b) 0842   (c) 2105.

*7 Change to seconds:
(a) 3 min 16 s   (b) 7 min 48 s.

8 Find the value of x if:
(a) $13x + 5(x + 4) - 9 = 4(2x + 8) - 1$
(b) $6(3x + 5) = 2(x - 5) + 4(2x + 6)$.

9 By substituting $y = x$ into the equation $2y + 3x = 15$ find a pair of values for x and y that satisfies (makes true) both equations. Illustrate your answer with a graph.

Paper Seven

1 Rewrite the following formulae so that r becomes the subject.
(a) $p = r - 4$ (b) $p = r + d$ (c) $m = t + r$ (d) $V = 4r$ (e) $k = pr - d$
(f) $d = try$.

2 Without using a calculator, and taking π as $\frac{22}{7}$, calculate for a circle of radius 21 cm:
(a) the circumference   (b) the area.

3 For Figure P10, without using a calculator and taking π as $3\frac{1}{7}$, calculate:
(a) the length of the arc
(b) the area of the sector.

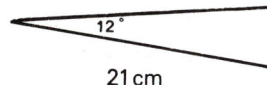

Fig. P10

4 Write as a decimal fraction:
(a) 7.3×10^{-2}
(b) the fraction displayed on a calculator as 6.04 $\quad-04$

5 Write in standard form ($A \times 10^x$ where $1 \leqslant A < 10$ and x is integral):
(a) 0.35   (b) 0.0035   (c) 706.

6 P = {a, b, r, u, p, t} and Q = {m, a, y, b, e}.

(a) Write the value of n(Q).   (b) List the elements of P ∩ Q.
(c) Illustrate sets P and Q with a Venn diagram.

*7 Using a calculator, and giving your answer correct to 3 s.f., simplify $\dfrac{375 \times 0.148}{16.35}$.

8 (a) 3 min 48 s + 51 s + 2 min 6 s   (b) 3 min 21 s − 1 min 43 s.

9 Multiply:
(a) $(4 \quad 2)\begin{pmatrix} 1 & -2 \\ -1 & 4 \end{pmatrix}$ (b) $\begin{pmatrix} 8 & 3 \\ 3 & 2 \end{pmatrix}\begin{pmatrix} 2 \\ -3 \end{pmatrix}$ (c) $\begin{pmatrix} 9 & 3 \\ 2 & 1 \end{pmatrix}\begin{pmatrix} 1 & -3 \\ -2 & 9 \end{pmatrix}$

10 Construct a triangle with sides 6 cm, 6 cm and 7 cm. Now construct the centre of the circumcircle and measure its radius.

Paper Eight

1 Use Pythagoras' Theorem to calculate side x, correct to 3 significant figures, in each triangle in Figure P11.

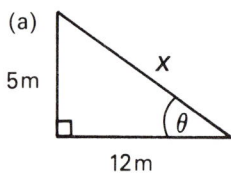

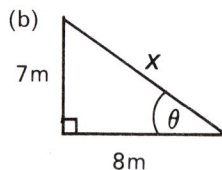

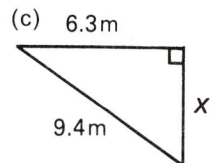

Fig. P11

2 Use the TAN key on your calculator to find x, correct to 3 significant figures, in Figure P12.

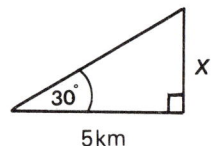

Fig. P12

3 Use the tangent ratio $\left(\dfrac{o}{a} = \tan\theta\right)$ to calculate the two angles marked θ in Figure P11. Give the answers to the nearest tenth of a degree.

4 Find the values of x and y that make the following pairs of equations true.

    (a) $2x + 4y = 16$
        $2x + 2y = 12$ SUBTRACT

    (b) $3x + 2y = 13$
        $2x - 2y = 2$ ADD

*5 Write as a 24-hour-clock time:
    (a) 6:30 a.m.   (b) 5:15 p.m.
    (c) ten minutes to eight in the morning
    (d) a quarter past nine in the evening.

6 Calculate, writing your answer in standard form:
    (a) $3.7 \times 10^4 \times 500$ (multiplication may be done in any order)
    (b) $7.8 \times 3 \times 2000$
    (c) $12.8 \times 100 + 4.2 \times 10^3$ (multiplication must be done before addition)

7 (a) Draw a travel graph to show the following journeys.

    Use 2 mm graph paper, with scales: 0 to 60 km, 1 cm represents 10 km; 8 a.m. to 7 p.m., 1 cm represents 1 h.

    Some cyclists left Epworth at 8 a.m. and cycled at an average speed of 20 km/h for $1\frac{1}{2}$ hours. They then stopped for 30 min before cycling for another 2 hours to a town 60 km from Epworth. An hour later they set off for Epworth, averaging 15 km/h until they reached the half-way point, where they stopped for half an hour. They completed the journey at an average speed of 10 km/h.

    Show also a second group of cyclists who left Epworth an hour later, joining the first group at 11 a.m. without a break.

    (b) (i) What was the average speed of the second group up to the time they joined the others?
        (ii) What was the average speed for the whole journey for each group, correct to the nearest km/h?

Paper Nine

1 Divide £35 in the ratio $2:3$.

2 (a) Increase 36 in the ratio $4:3$.

    (b) Decrease 24 in the ratio $5:16$.

3 A map is made to the scale $1:25\,000$.

    (a) How many metres are represented by: (i) 1 cm (ii) 2.3 cm?

    (b) What distance on the map would represent 500 metres?

4 The sill of a window is 4.8 metres above the ground. What is the length of the shortest ladder that can just reach the sill if the base of the ladder has to be at least 2 metres from the wall?

5 Use tangents to find θ and x in Figure P13.

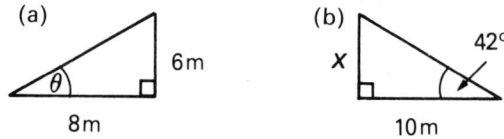

(a)

6m

8m

θ

(b)

42°

x

10m

Fig. P13

6 Express correct to 3 significant figures:
(a) 2.324 (b) 0.03876

7 The cost of buying some books valued at £2.35 each, and of having them delivered, was £42.31. The delivery charge was £2.36. How many books were bought?

***8** Copy Figure P14, using a radius of 3cm for the complete circle.

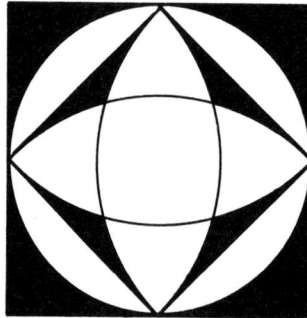

Fig. P14

9 Using the ratio of the sides of similar triangles, calculate the unknown sides in Figure P15.

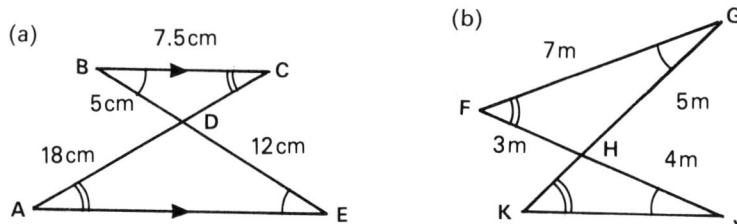

(a)

7.5cm

B   C

5cm

D

18cm   12cm

A   E

(b)

G

7m

F

5m

3m

H

4m

K   J

Fig. P15

10 Make a the subject of:
(a) $P = 2a + 2b$ (b) $r = 2a + p$ (c) $V = ab(R - r)$.

11 A pond of radius 12m is to be surrounded by a lawn 4m wide. Taking $\pi = 3.142$, and using a calculator, find the area of the lawn.

Paper Ten

1   A bag contains 4 black, 1 white, and 3 grey counters. State as the simplest possible fraction the probability of picking, if each picked counter is immediately replaced:
(a) a black   (b) a white   (c) a black then a white   (d) two blacks
(e) a black and a grey in either order.

2   Write as a 24-hour-clock time:
(a) 6:30 a.m.   (b) 5:15 p.m.

3   Give the number of hours and minutes between:
(a) 7:14 a.m. and 8:06 a.m.   (b) 7:42 a.m. and 11:37 a.m.   (c) midday and 4:32 p.m.

4   Find the value of y if:
(a) $4(2y - 1) = 20$ (b) $9(y + 1) - 6 = 2(y + 12)$.

5   (a) Increase 35 cm in the ratio $6 : 5$.
    (b) Decrease £48 in the ratio $5 : 6$.

6   In how many different ways (that is, so that at least one man has at least one different person by his side) can four men:
(a) sit on a park bench   (b) sit round a table?

*7   Use Pythagoras' Theorem to calculate side h in Figure P16.

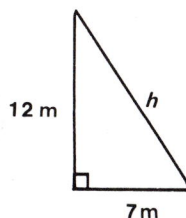

Fig. P16

8   The average age of the eleven players in a hockey team is 19 years 7 months. If the oldest player is 22 years 11 months old, what is the average age of the rest of the team?

9   In this question check each answer carefully, as one error will lead to all the answers afterwards being wrong. Your final answer, A^6, should be $\begin{pmatrix} -1 & 0 \\ 0 & -1 \end{pmatrix}$.

$A = \begin{pmatrix} 0 & 1 \\ -1 & 0 \end{pmatrix}$. Find:

(a) A^2 (that is, $A \times A$)   (b) A^3 (that is, $A \times A^2$)   (c) A^4 (d) A^5 (e) A^6.

Now draw a pair of axes, each from -4 to 4, and show the result of multiplying the matrix $\begin{pmatrix} 0 & 3 & 3 \\ 0 & 0 & 2 \end{pmatrix}$, which describes a triangle, by A, A^2, A^3, A^4, A^5 and A^6. Describe fully each transformation.

Paper Eleven

1 (a) Subtract 3.96 from 8.007
(b) Add two hundredths to 125.699

2 (1) $2\frac{1}{4} + 1\frac{1}{3}$ (b) $2\frac{1}{4} - 1\frac{1}{3}$ (c) $2\frac{1}{4} \times 1\frac{1}{3}$ (d) $2\frac{1}{4} \div 1\frac{1}{3}$

3 **Example** $48 = 2 \times 2 \times 2 \times 2 \times 3$

Write as a product of prime factors:
(a) 12   (b) 16   (c) 28   (d) 36   (e) 60.

4 Find the highest common factor (HCF) of:
(a) 16 and 24   (b) 28 and 35   (c) 27 and 30.

5 Find the lowest common multiple (LCM) of:
(a) 4 and 6   (b) 8 and 12   (c) 9 and 15.

6 Use a ruler and compasses, but not a protractor, in this question.

(a) Construct $\triangle ABC$ where $\angle A = 90°$, AB = 4.5 cm and AC = 6 cm. Measure BC.

(b) Construct $\triangle DEF$ where $\angle D = 90°$, $\angle E = 60°$ and DE = 4 cm. Measure EF.

*7 Two sides of a right-angled triangle are 12 cm and 16 cm. Use Pythagoras' Theorem to calculate the length of the hypotenuse.

*8 The scale of a map is 5 cm represents 1 km. What distance is represented by:
(a) 1 cm   (b) 4.2 cm?

*9 Draw a circle of radius 45 mm. Divide it accurately into twelve equal sectors.

10 Evaluate, without using a calculator and showing all necessary working:
(a) $3.8 + 0.048 + 24$ (b) $194 - 0.37$ (c) $6.2 - 0.57$ (d) $(0.2)^2$ (e) $\sqrt{0.16}$
(f) $1.6188 \div 0.38$

11 Copy Figure P17, then shade the region $A \cap B'$.

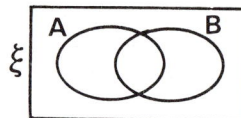

Fig. P17

Paper Twelve

1 $A = \begin{pmatrix} 4 & 1 \\ 1 & -4 \end{pmatrix}$; $B = \begin{pmatrix} 0 & 3 \\ 3 & 5 \end{pmatrix}$

Calculate:
(a) $A + B$ (b) AB (c) BA (d) A^2 (e) A^{-1} (the inverse of A).

2 Solve by the matrix method the simultaneous equations
$3x + 2y = 12$ and $x + y = 3$.

3 (a) $2\frac{2}{3} + 3\frac{2}{5}$ (b) $3\frac{1}{4} - 2\frac{2}{3}$ (c) $5\frac{3}{5} \times 3\frac{4}{7}$ (d) $3\frac{3}{4} \div 2\frac{1}{3}$

4 (a) $5\,000\,000 \times 5\,000\,000$
   (b) Find a quarter of three-eighths.

5 Write as a decimal fraction:
   (a) $\frac{1}{2}$ (b) $\frac{1}{4}$ (c) $\frac{3}{4}$ (d) $\frac{1}{10}$ (e) $\frac{1}{100}$.

6 A box holds $1000\,cm^3$ of powder and has a width of 11 cm. Calculate the width of a box, mathematically similar in shape, but holding $27\,000\,cm^3$ of powder. (Note that the answer is not 287 cm!! Remember the connection between lengths' ratio and volumes' ratio.)

7 (a) In Figure P18, find the third angle, and use this angle and the tangent ratio to calculate side x correct to 3 significant figures.

   (b) Why is it easier to use the third angle rather than the 48° one?

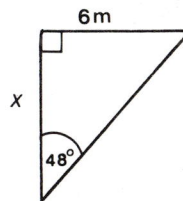

Fig. P18

*8 A bag contains 5 red, 2 white, and 3 blue counters. State the probability of picking, if the first counter picked is replaced, first a red and then a white.

9 Showing all working calculate:
   (a) 76×80 (b) 9.3×0.54 (c) $10.2 \div 0.2$

10 Design a diagram to show the six possible ways that Alan (A) and Bob (B) could each choose a dancing partner from Clare (C), Dawn (D) and Eve (E) if Alan always asks a girl to dance before Bob plucks up the courage.

Use your diagram to find the probability that:
(a) Alan asks Eve to dance
(b) Alan does not ask Eve to dance
(c) Bob asks Eve to dance
(d) Dawn dances with one of the two lads
(e) Dawn does not dance with one of the two lads
(f) Clare dances with one of them, but Eve does not.

Paper Thirteen

1 In Figure P19, O is the centre of the circle.
Calculate, with reasons:
(a) angle a (b) angle b.

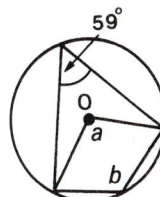

Fig. P19

P

2 For Figure P20, taking $\pi = \frac{22}{7}$:
  (a) calculate the length of the arc
  (b) find the area of the sector.

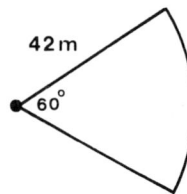

Fig. P20

3 A bag contains 4 green (G), 1 red (R) and 3 yellow (Y) balls. State as a simplified fraction the probability of picking, if each picked ball is replaced:
  (a) GR (green then red)   (b) RR   (c) GRY   (d) RGG.

4 Multiply: $\begin{pmatrix} 4 & 2 \\ 1 & 1 \end{pmatrix} \begin{pmatrix} 1 & -2 \\ -1 & 4 \end{pmatrix}$.

5 Write as a product of prime factors:
  (a) 12   (b) 15   (c) 38   (d) 48.

6 State the highest common factor of:
  (a) 12 and 16   (b) 32 and 68   (c) 24, 16 and 56.

7 State the lowest common multiple of:
  (a) 12, 16 and 30   (b) 15, 18 and 30.

*8 Simplify:
  (a) $8\frac{1}{4} - 5\frac{2}{5}$ (b) $1\frac{5}{6} \times 1\frac{1}{8}$.

*9 What is the longest straight-line race-track that can be laid out in a rectangular field 80 metres long and 60 metres wide?

10 A ship sails 9 miles north then 17 miles west then 9 miles north again. Calculate how far and on what bearing it then is from its starting position.

11 In Figure P21, O is the centre of the circle. Calculate:
  (a) the perpendicular distance from O to AB
  (b) the length of the chord CD.

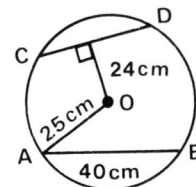

Fig. P21

12 Simplify:
  (a) $(\frac{4}{5} - \frac{3}{4}) \div \frac{7}{8}$ (b) $\frac{1}{4}(\frac{5}{6} + 1\frac{2}{3}) \div \frac{7}{12}$

196

Paper Fourteen

1 Calculate the angle θ in both triangles in Figure P22.

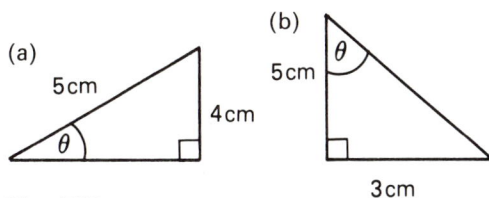

Fig. P22

2 Simplify:
(a) $(37.2 \div 6) - (0.34 \times 6)$ (b) $(3.5 \times 8) + (1.33 \times 6)$.

*3 Find the highest common factor of:
(a) 32 and 48   (b) 24, 42 and 51.

*4 How many 7p stamps can be bought for £3.99?

*5 A rectangular carpet is 7 m long and 4 m wide. Calculate:
(a) the area of the carpet   (b) the cost of the carpet if 1 m^2 costs £7.20.

*6 A car travels on a motorway at a constant speed of 120 km/h. How far does it travel in:
(a) 2 hours   (b) half an hour   (c) 20 minutes?

*7 Solve:
(a) $27 + 3x = 5x + 15$ (b) $5x + 4 = 2x + 10$.

8 Simplify:
(a) $(\frac{1}{3} - \frac{1}{4}) \div (\frac{1}{2} - \frac{1}{3})$ (b) $1\frac{1}{4} \div (6\frac{1}{3} - 2\frac{7}{12})$ (c) $(2\frac{1}{2} \div 3\frac{1}{3}) + (1\frac{1}{2} \times 2\frac{1}{3})$
(d) $\frac{3}{5}$ of $\frac{15}{24} + (2\frac{5}{6} - 1\frac{1}{2})$.

9 Giving the answer in standard form, find:
(a) $8 \times 10^4 \times 3.6 \times 10^2$ (b) $(7.45 \times 10^{-2}) + (3.6 \times 10^{-1})$.

10 Solve:
(a) $2(4a + 3) - 4a = 2(a - 1) + 14$ (b) $7(b - 2) = 3(3b + 4)$.

11 Solve by the matrix method:
$8x - 5y = 7$ and $7x - 3y = 13$.

12 Sixty members of an athletics club attempted to gain standards in three events. If 26 gained a standard in running, 30 in throwing, 12 in running and jumping, 16 in jumping and throwing, and 4 in running and throwing, while 6 gained none, how many gained:
(a) 3 standards   (b) only 2 standards   (c) only 1 standard?

Paper Fifteen

1 Multiply:
(a) $(x + 3)(x + 2)$ (b) $(x - 4)(x - 2)$.

2 Solve simultaneously: $2x + y = 4$
$3x + 2y = 5$

3 Give the number of hours and minutes between:
(a) 7:14 a.m. and 1:20 p.m.   (b) 7:14 a.m. and 9:06 p.m.   (c) 9:38 p.m. and 1:27 a.m.

4 (a) $83.5 + 5.248 + 27$ (b) $(1\frac{1}{8} \times 4) + (2\frac{5}{6} \times 3)$

5 Oranges on Mike's market stall are priced at 7 for 24p. On Mary's they are 42p a dozen. Showing clear working, find who is selling the cheaper oranges.

*6 Accurately construct Figure P23, making the large circle of radius 5 cm.

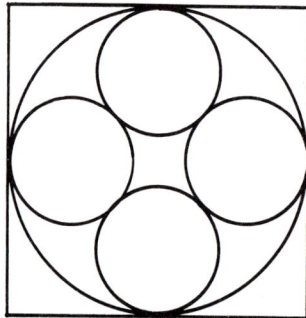

Fig. P23

*7 Write down the next term in the sequences:
(a) 3, 7, 11   (b) 2, −2, −6   (c) 1, 3, 6, 10.

8 A cable-railway is 4400 m long and ascends an average slope of 35°. What height does it rise to, to the nearest metre?

9 The scale of a map is 1 : 50 000.

(a) Find the distance in km represented by 30 cm on the map.

(b) Find the true area in km² of a lake of area 2 cm² on the map.

10 A circular saw of diameter 40 cm is driven at 1500 revolutions per minute. What is the speed of the cutting edge in metres per second, correct to the nearest whole number? (Take $\pi = 3.142$. You may use a calculator.)

Paper Sixteen

1 State the co-ordinates of the crossing point of:
(a) $x = -1$ and $y = 0$ (b) $y = x$ and $x = 5$
(c) $y = 2$ and $y = x$ (d) $y = x + 1$ and $x = 2$.

2 Complete the co-ordinates $(2, \quad)$, $(0, \quad)$ and $(-4, \quad)$ to lie on the line:
(a) $y = x$ (b) $y = 3x$ (c) $y = -2x$.

3 A ship sails 20 km due south and then 14 km due west. Calculate how far it then is from its starting point, correct to 3 s.f., and its bearing from this point, correct to one-tenth of a degree, both as a three-figure and as a cardinal bearing.

4 Using a ruler and compasses, but not a protractor:

(a) construct $\triangle ABC$ where AB = 8 cm, $\angle A = 75°$ and AC = 6 cm, then state the length of BC

(b) construct $\triangle DEF$ where EF = 6 cm, $\angle F = 60°$ and $\angle E = 75°$, then state the length of DE.

5 If $v = u + at$:
(a) rewrite the formula to make u the subject
(b) calculate u when $v = 16$, $a = 3.5$ and $t = 4$.

*6 Simplify $(2\frac{1}{2} - 1\frac{3}{4}) \times (2\frac{3}{4} + 1\frac{1}{2})$.

*7 Write as a denary number:
(a) 1.6×10^4 (b) 9.6×10^{-2}.

*8 Write in standard form, as in question 7:
(a) 136   (b) 0.0097

9 Simplify:
(a) $(5\frac{1}{3} \div 1\frac{3}{5}) + (1\frac{2}{3} \times \frac{7}{10})$ (b) $4\frac{2}{5} \div (\frac{1}{3} + \frac{2}{5})$.

10 Solve simultaneously by substitution:
$y = 2x + 8$ and $2y + x = 1$.

11 Giving your answer in standard form, simplify:
(a) $4.3 \times 10^{-2} \times 2 \times 10^4$ (b) $6.7 \times 10^2 + 3.4 \times 10^{-2}$.

Paper Seventeen

1 Factorise:
(a) $2a + 8$ (b) $b^2 - 4$ (c) $c^2 - 4c$.

2 Calculate angles a to e in Figure P24.

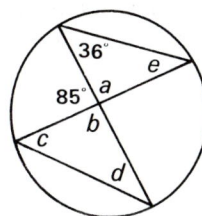

Fig. P24

3 Calculate the lengths x, correct to 2 s.f., in Figure P25.

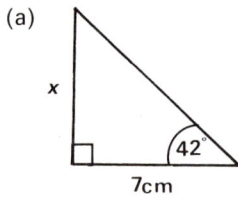

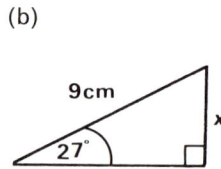

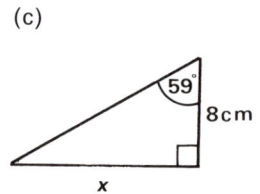

(a)

(b)

(c)

Fig. P25

4 A bag contains 5 red, 2 white, and 3 blue counters. State the probability of picking, if the first counter is not replaced before the second one is picked:
(a) red then white   (b) red then red.

5 In question 4, which of the two events is the more likely?

6 Calculate as a fraction in its simplest form:
(a) $\frac{1}{3}$ of $2\frac{3}{4}$ (b) $\frac{7}{8} - \frac{1}{3}$.

7 The average age of a form of 20 pupils is 14 years 9 months. Find the total of their ages in years.

*8 A car mileage meter reads 56 716 when the tank is filled. It reads 56 909 when it is next filled, the tank then taking 25 litres of petrol. What is the average fuel consumption in miles per litre, correct to the nearest whole number?

*9 Posts cost £8 each and six-foot fencing panels cost £15 each.

(a) How many posts and panels to fence 32 feet of garden? (The posts are 4 inches (a third of a foot) square.)

(b) What is the cost of fencing the garden?

10 Solve simultaneously $y = x$ and $y = 2x - 4$ by a graphical method, using axes x from 0 to 4, y from -4 to 4.

11 A pendulum swings through an arc of 12°. If the tip of the pendulum is 21 cm from the point of suspension, how long is the arc it makes? (Take $\pi = 3\frac{1}{7}$.)

12 In question 11, through what area does the pendulum sweep?

13 Draw a picture to show how a pole should be held to cast the longest possible shadow at about ten o'clock one sunny morning.

Paper Eighteen

1 Simplify:
(a) $a^4 \times a^2$ (b) $4a^2 \times 3a^3$ (c) $a^6 \div a^4$ (d) $33a^4 \div 3a^3$.

2 (a) $V = a^3$. Find V when $a = 3$.

(b) $A = \frac{1}{2}bh$. Find A when $b = 16$ and $h = 3.6$.

(c) Where have you met the formulae in (a) and (b) before?

3 State the sizes of the marked angles in Figure P26, in alphabetical order with reasons.

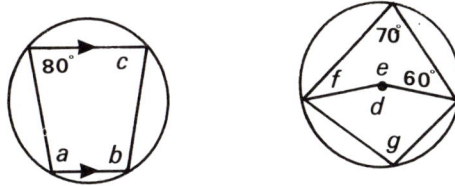

Fig. P26

4 A girl cycles 20 km in an hour and a boy cycles 5 m in a second. Who goes the faster, and by how many km h^{-1}?

5 Express correct to 1 significant figure:
(a) 7284   (b) 3.7265   (c) 0.026 17

6 Repeat question 5, but correct to 2 significant figures.

7 Repeat question 5, but correct to 3 significant figures.

8 Given that A = $\begin{pmatrix} 1 & 0 \\ 0 & -1 \end{pmatrix}$ and B = $\begin{pmatrix} 3 & -1 \\ 5 & 2 \end{pmatrix}$, calculate:

(a) A + 2B   (b) AB   (c) B^{-1} (the inverse of B).

*9 Write one addition, one subtraction, one multiplication, and one division, to which the answer is:
(a) 25   (b) 6.3

*10 Find the value of the letter if:

(a) $\frac{x}{5} = 10$ (b) $4 - a = 1$ (c) $1 - y = 4$ (d) $2k - 3 = 9$.

11 Expand:
(a) $(2x - 4)(3x + 2)$ (b) $(x - 4)(3x - 2)$.

12 Find the size of angles x and y in Figure P27.

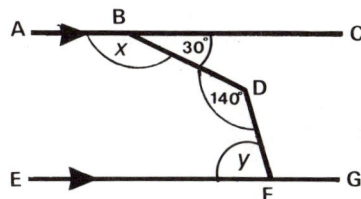

Fig. P27

13 For the four integers n, $n + 1$, $n + 2$, and $n + 3$, say which of the following are (a) always true, (b) sometimes true, (c) never true.

A: Two are even   B: Two are multiples of 3   C: One is a multiple of 4
D: Their sum is exactly divisible by 4   E: Their product is exactly divisible by 120
F: The sum of two of them is half their total sum
G: The product of two of them equals the product of the other two.

Paper Nineteen

1 Express:
(a) 16 out of 40 as a percentage
(b) 250 g as a percentage of 1 kg.

2 Jethro buys an antique chair for £400 and sells it at a gain of 45%. What is his selling price?

3 In a year Polly grows from 1.50 m to 1.55 m. What is the percentage change in her height? (Work in centimetres.)

4 Simplify: $4a^2 - 3ac + 8ac - 2a^2 + 14$.

5 Using a calculator, and taking π as 3.142, calculate C, A and V, correct to 2 decimal places, if:

(a) $C = 2\pi R$ and $R = 3.6$ (b) $A = \pi r^2$ and $r = 2.7$ (c) $V = \dfrac{4\pi r^3}{3}$ and $r = 6.2$

6 In Figure P28, O is the centre of the circle. Calculate:
(a) angle a (b) angle b.

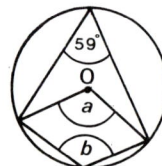

Fig. P28

***7** Copy Figure P29, making the large square of side 2.5 cm. Then fill in further squares to give it rotational symmetry of order 4.

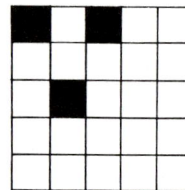

Fig. P29

8 How many complete turns does a car wheel of diameter 24 inches make in a mile? (12 inches = 1 foot; 3 feet = 1 yard; 1760 yards = 1 mile. Take $\pi = 3.14$.)

9 An object travelling initially at a velocity of u m/s and accelerating at a m/s^2, will travel s metres in t seconds, where $s = ut + \frac{1}{2}at^2$, assuming that there is no resistance to the motion.

Find the distance travelled by an object in 3 seconds if it is travelling initially at 80 m/s, and is accelerating at 32 m/s^2.

10 Simplify:

(a) $\dfrac{3ab \times 16a^3b}{12ab^2}$ (b) $2ab^2 \times a^3b \div 4a^2b$.

11 Find the simultaneous solution of $y = 2x - 4$ and $y = x - 1$ by drawing a graph on axes: x from 0 to 4, y from -4 to 4.

Paper Twenty

1 Copy and complete the table for the graph of $y = x^2 - 3x - 4$, then draw the parabola, using 1 cm to 1 unit on both axes.

x	-2	-1	0	1	2	3	4	5
x^2		1	0		4		16	25
$-3x$		3	0		-6		-12	-15
-4		-4	-4		-4		-4	-4
y		0	-4		-6		0	6

2 The price of a £20 radio rises by 18%. Find the new price.

3 Factorise:
(a) $4a - 8$ (b) $a^2 - 9$ (c) $3c^3 + 2c^2 + c$

4 Showing all working, calculate:
(a) 2.35×1.32 (b) $0.56 \div 0.7$

*5 Calculate:
(a) $\frac{1}{3} - \frac{1}{4}$ (b) $2\frac{1}{2} + 5\frac{1}{3}$ (c) $6 - 3\frac{7}{8}$

*6 A ship sails 5 nautical miles east, then 12 nautical miles north. Calculate how far it then is from its starting point.

7 Calculate:
(a) $1\frac{7}{8} \times 3\frac{1}{5} \div \frac{4}{9}$ (b) $3\frac{1}{3} \times 2\frac{1}{4} - 1\frac{1}{2}$

8 Solve simultaneously: $3x + 6y = 12$ and $2x + 5y = 11$.

9 Find the original cost of an item priced £24 in a '20% off' sale.

10 Three prizes together cost £48. The second is worth £7 more than the third and £10 less than the first. What is each prize?

P

11 Figure P30 shows the path of a windscreen wiper. What area of glass is swept clear by the blade? (Use $\pi = \frac{22}{7}$.)

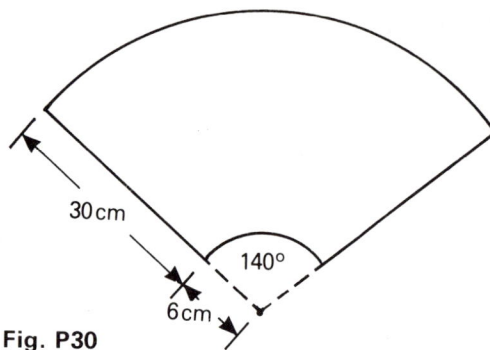

Fig. P30

1 State the sizes of angles *a* to *e* in Figure P31.

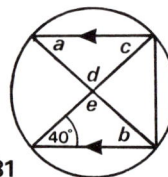

Fig. P31

2 Showing all working, simplify:
(a) $3.8 + 0.048 + 24$ (b) $194 - 0.37$ (c) $6.2 - 0.57$ (d) 0.2×0.2
(e) $9.684 \div 12$ (f) 147.5×11.2

3 A television originally costing £280 is reduced in a sale by 15%. What is the new price?

*4 Using the British Airways information given below, calculate the cost of booking in March a one-way flight departing on Saturday August 1st, for two adults and one 18-month old infant going on a fortnight's holiday.

SCHEDULED SERVICES			
British airways			
Days of departure:	Daily to J.F.K.: dep 11.00, arr 13.30; dep 13.15, arr 15.45; dep 18.30, arr 21.00		
Aircraft:	Boeing 747/TriStar		
Fares:	Super Apex (return)	Economy (one-way)	Standby (one-way)
April 1–May 31 and Oct 1–Oct 31	£329 (Mon–Thu) £369 (Fri–Sun)	£233 (Mon–Thu) £248 (Fri–Sun)	£170 (available for travel from July 1)
June 1–Sept 30	£349 (Mon–Thu) £389 (Fri–Sun)	£267 (Mon–Thu) £282 (Fri–Sun)	
Child discount:	Infants under 2 years 90%; 2–11 years 33.3%		Infants 90%
Conditions:	Must be booked at least 21 days before travel. Min stay 7 days, max stay 6 months.	—	Must travel within 3 months of purchase. Seat confirmed at airport. No standbys on eastbound flights after Sept 30.
Comments:	Pay bar. Charge for movies/stereo.		

5 In Figure P32, AB = 10 cm, BC = 24 cm and AD = 6 cm.
Calculate:
(a) the length of DB
(b) the length of CD
(c) the area of triangle ABC.

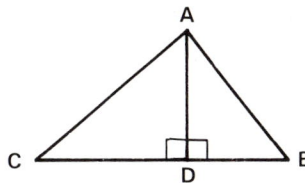

Fig. P32

6 Make r the subject of:
(a) $r - p = 9$ (b) $p = rt$ (c) $c = 2\pi r$.

7 Remembering to cancel as much as possible, simplify:
(a) $1\frac{1}{2} \times 1\frac{1}{3} \times 1\frac{1}{4}$ (b) $1\frac{1}{2} \times 1\frac{1}{3} \times 1\frac{1}{4} \times 1\frac{1}{5}$ (c) $1\frac{1}{2} \times 1\frac{1}{3} \times 1\frac{1}{4} \times 1\frac{1}{5} \times 1\frac{1}{6}$

8 On axes: x from -4 to 4 and y from 0 to 9, plot the graphs of $3x + 2y = 9$ and $y - 3x = 9$, and hence solve these two equations simultaneously.

Also shade on your graph the region
$\{(x, y): 3x + 2y \geqslant 9; y - 3x \leqslant 9; x \leqslant 0\}$

9 Simplify:
(a) $\dfrac{42a^4 b^3}{36a^2 b}$ (b) $15a^6 \div 3a^4$ (c) $16a^5 b^4 \div 4a^5 b^2$ (d) $4a^6 \times 4a^3 \div 8a^4$.

Paper Twenty-two

1 Find for a regular octagon:
(a) the size of one exterior angle
(b) the size of one interior angle
(c) the sum of the interior angles.

2 Find the interior angle sum of a polygon with 39 sides.

3 The population of a town was 400 000 in 1969 and 464 000 in 1970. What was the increase percent?

4 Solve simultaneously: $5x - 2y = 13$
$3x + y = 21$

5 Express correct to one significant figure:
(a) 43 527   (b) 5.007 34   (c) 0.038 37

6 Write the numbers in question 5 correct to two significant figures.

7 Simplify:
(a) $a^3 \times a^2$ (b) $4a^2 \times 3a^3$ (c) $b^3 \times 2a^2 b$.

*8 Find the size of each angle in Figure P33.

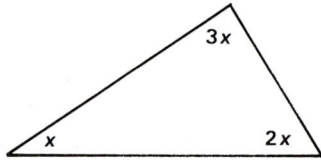

Fig. P33

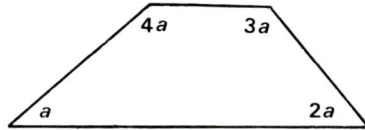

*9 Copy Figure P34 as neatly as you can. Make the sides of the large triangle about 60 mm.

Fig. P34

10 Using two sets of axes, from −5 to 5 each, hatch *outside* the following regions.
(a) $\{(x, y): -3 \leqslant x < 1; 2 < y < 4\}$
(b) $\{(x, y): x + 1 \geqslant y > 1; x \leqslant 3\}$

11 Alexa sells a stamp for £56. The number of pounds that she paid for it is the same as the percentage profit made when she sold it. What did she pay for it?

12 On a journey of 143 miles a train covered the first 60 miles at 50 m.p.h. and the next 40 miles at 30 m.p.h. At what rate was the remainder of the journey covered if the average speed for the whole journey was 40 m.p.h.? (You should use a calculator.)

Paper Twenty-three

1 Find the mean of 36, 42, 37, 41, 35, 40, 36, 37, 40 and 39.

2 Use your answer to question 1 to write down the mean of 236, 242, 237, 241, 235, 240, 236, 237, 240 and 239.

3 When throwing rings at a target Alison obtains the following scores:

Score	0	1	2	3	4	5
Number of rings with this score	15	20	16	15	24	10

Find:
(a) the modal score   (b) the median score   (c) the mean score.

4   Calculate the size of each lettered angle in Figure P35.

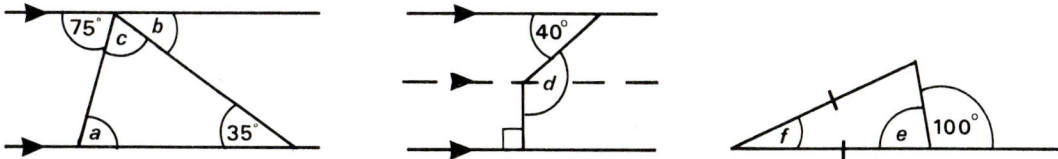

Fig. P35

5   How many sides has a regular polygon if its exterior angles are each:
(a) 36°   (b) 20°   (c) $22\frac{1}{2}$°?

*6   Cary earns £75 and spends $\frac{3}{5}$ of it. How much has he left?

*7   How many boxes of ten 30 cm square carpet tiles should be bought to carpet a rectangular area 9 metres long and 6 metres wide?

8   In Figure P36 state the size of:
(a) ∠ABC   (b) ∠ACB   (c) ∠BAC
(d) ∠ADC   (e) ∠BDC   (f) ∠ABD
(g) ∠CBD   (h) ∠CAD.

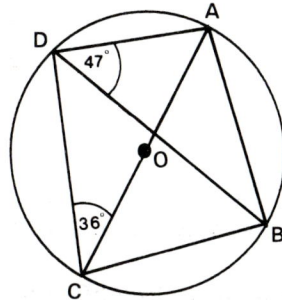

O is the centre

Fig. P36

9   The price of a ticket rises from £6.25 to £7.25. What is the increase percent?

10   In its first 8 matches a hockey team averaged 3.25 goals per match. After 12 matches the average was 3.5 goals per match. How many goals were scored in the last four matches?

11   Andrew's salary is £10 000. His tax-free allowance is £2 800. He is taxed on the remainder at 30%. He also pays 6% of his gross salary to a superannuation (pension) fund, and National Insurance of 8% on the first £1 500 of his gross salary and 5% on the remainder. What is his monthly net income?

Paper Twenty-four

1   Factorise:
(a) $3x + 30$ (b) $4x^2 - 1$ (c) $4c^3 + 6c^2$.

2   Write down the value of h if $\dfrac{25}{30} = \dfrac{5}{h}$.

3 Calculate the values of x, y, θ, α and z in Figure P37.

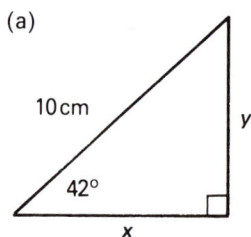

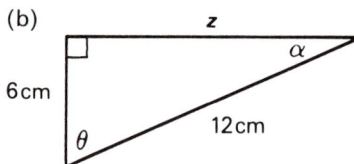

(a)

10 cm
y
42°
x

(b)

6 cm
z
α
12 cm
θ

Fig. P37

4 Write 12 cm² in m². (It is *not* 0.12 m².)

*5 A typist types 24 pages, each containing an average of 270 words, in one hour. How many words could she type:
(a) in 1 hour   (b) in 6 hours   (c) in 1 minute?

*6 Write the next term in the sequence:
(a) -3, 6, -12 (b) 2, 3, 5, 9, 17   (c) $-\frac{1}{2}$, $\frac{1}{4}$, $-\frac{1}{8}$.

*7 If $a = 2$, $b = -3$, and $c = 4$, find the value of:
(a) abc (b) $c^2 - b$ (c) $2c^2$ (d) a^3.

*8 Find the value of the letter if:
(a) $9 - a = 6$ (b) $a - 9 = 6$ (c) $6a = 9$ (d) $9a = 6$ (e) $\frac{a}{9} = 6$.

9 Given that $a = \frac{1}{3}$ and $b = -2$, evaluate:
(a) $6a - 2b$ (b) $9a^2 - 2b^2$.

10 Find the simultaneous solution of $y = x - 3$ and $2y = x - 1$ by a graphical method. Use axes: x from 0 to 7, y from -3 to 3.

11 Make a the subject of:
(a) $s = ut + \frac{1}{2}at^2$ (b) $h = \dfrac{24b}{a + b}$.

12 For $\triangle ABC$ in Figure P38 calculate:
(a) its area   (b) the length BC
(c) the perpendicular distance of A from BC
(d) $\angle ABC$ and $\angle ACB$ correct to the nearest 0.1°.

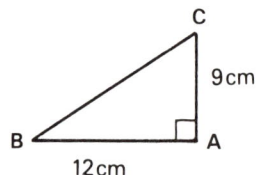

C
9 cm
B
12 cm
A

Fig. P38

13 Work out

$$(7.028 \times \sqrt{5.3}) - \frac{300\pi - \sin 25°}{\sqrt[3]{30} + 1} + \frac{3 \times 10^3}{16.9 \times 0.18}$$

correct to 3 s.f. (You may use a calculator!)

Paper Twenty-five

1 In Figure P39, O is the centre and ATB is a tangent.
AT = OT. Calculate the sizes of angles a to d.

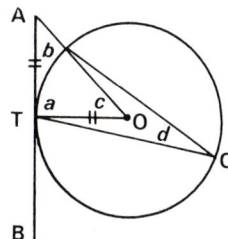

Fig. P39

2 In Figure P40 XOY is the diameter of the circle, centre O.

(a) State the sizes of angles a and b.

(b) Use trigonometry to calculate angle c.

(c) Calculate the shortest distance from O to XB.

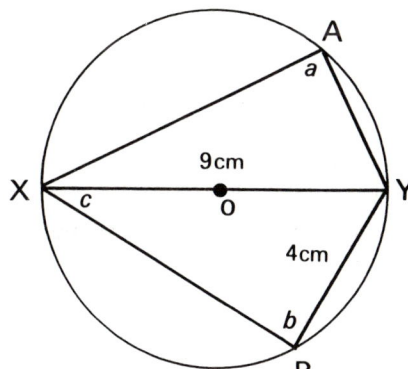

Fig. P40

3 Factorise $x^2 + 7x + 10$.

4 Simplify:
(a) $(x - 4)(x - 2)$ (b) $(x + 2)(3x - 4)$.

5 (a) Copy and complete the table, then draw the graph of $y = x^2 + 2x - 4$.
Use scales: x, 2 cm to 1 unit: y, 1 cm to 1 unit.

x	-5	-4	-3	-2	-1	0	1	2
x^2	25		9	4			1	
$+2x$	-10		-6	-4			2	
-4	-4		-4	-4			-4	
y	11		-1	-4			-1	

(b) State the values of x that make $x^2 + 2x - 4 = 0$, giving your answers correct to one place of decimals.

*6 Simplify:
(a) $\frac{3}{4} + \frac{2}{3} - \frac{1}{2}$ (b) $5\frac{4}{5} - 3\frac{1}{10}$ (c) $3\frac{1}{3} - 1\frac{3}{4}$ (d) $\frac{2}{3} \div \frac{5}{6}$ (e) $1\frac{1}{2} \div 1\frac{7}{8}$.

7 Evaluate: $\frac{5}{6}(2\frac{7}{8} + 1\frac{1}{2}) \div 2\frac{3}{16}$.

8 Write $3\,cm^2$ in m^2.

9 The mean average age of 15 girls was 14 years 7 months. When one girl was away the mean average age fell to 14 years 6 months. How old was the absent girl?

10 Write without indices:

(a) 4^3 (b) $8^{\frac{2}{3}}$.

11 In Figure P41, use the sine rule to help you find:

(a) $\angle A$ (b) $\angle C$ (c) AB.

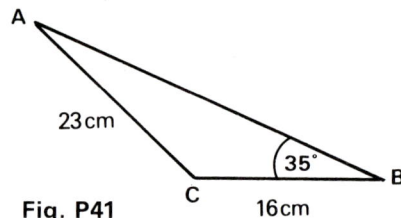

Fig. P41

Paper Twenty-six

1 Calculate the sizes of the lettered angles in Figure P42. O is the centre of the circle.

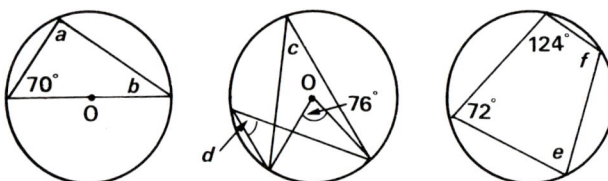

Fig. P42

2 In Figure P43 calculate:

(a) BC (use $\angle A$)   (b) CD   (c) BD.

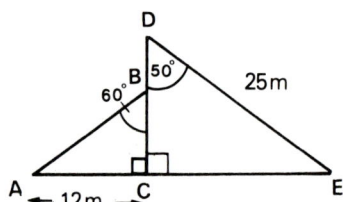

Fig. P43

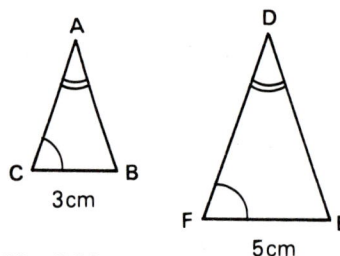

Fig. P44

3 Figure P44 shows two similar triangles. Triangle ABC has an area of $9\,cm^2$. Calculate:

(a) the area of triangle DEF

(b) the length of the altitude from A to CB.

4 Factorise:

(a) $2x^2 + 4x$ (b) $4a^2 - 25$ (difference of two squares)

(c) $2ab + 4a^2 + 6ab^2$.

5 (a) Write $2\,m^2$ in cm^2.   (b) Write $200\,cm^2$ in m^2.

*6 An examination is timed to take $1\frac{3}{4}$ hours. At what 24-hour-clock time should it finish if it starts at:

(a) 9 a.m. (b) 9:40 a.m. (c) 2:12 p.m.?

*7 How many days are there from 18 March to 12 June inclusive?

*8 Use a calculator to find $\dfrac{130 \times 24.6}{17 \times 93}$ correct to 3 s.f.

9 Write without indices:

(a) 5^{-1} (b) 2^{-3} (c) $16^{\frac{3}{4}}$ (d) $27^{\frac{2}{3}}$.

10 Use a calculator to find a, correct to 3 s.f., if $a^3 = \dfrac{3}{0.071\,36}$.

11 Given that $x = \dfrac{1 + z}{z}$ express z in terms of x.

(Note: You cannot use the flow-diagram method, as the subject letter appears twice in the formula.)

12 Solve:

(a) $3(x - 2) - 2(3x - 3) + 1 = 0$ (b) $\dfrac{x}{3} + \dfrac{x}{2} = \dfrac{5}{6}$.

13 Mary is 8 years older than Nicola. In 5 years' time she will be twice as old as Nicola is then. How old are they now?

Paper Twenty-seven

1 A solid cylinder has a diameter of 14 cm and is 11 cm high. Find:
(a) the area of one end (b) the curved surface area.
(Take $\pi = \frac{22}{7}$.)

2 The outside measurements of a box without a lid are: length 32 cm, breadth 27 cm, depth 10 cm. If it is made from 1 cm thick plastic find the inside volume.

3 Figure P45 shows the path of a boat which sails 8 km from port A to a point B on a bearing of 220°. Calculate the distances BC and AC (called its 'northing' and 'easting' from A).

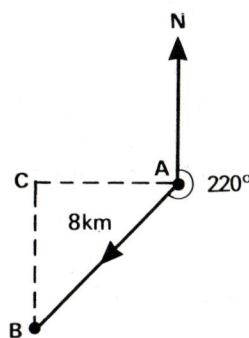

Fig. P45

4 Find for 56, 14 and 28:
(a) their H.C.F.   (b) their L.C.M.

*5 Draw and label two pairs of axes from −5 to 5 each. On one set of axes mark the points (−3, 1) and (−3, 4). On the other set mark the points (2, 3) and (−4, 3). State the equations of the straight lines that join your pairs of points.

*6 Using a pencil, compasses and a ruler, draw Figure P46, giving the circle a radius of 3 cm.

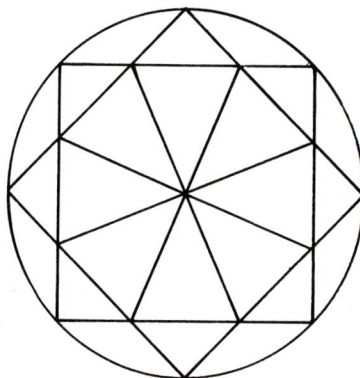

Fig. P46

7 Expand:
(a) $(2x - 4)(x - 2)$ (b) $2(3a + 4)(2a - 2)$.

8 Calculate angles x and y in Figure P47.

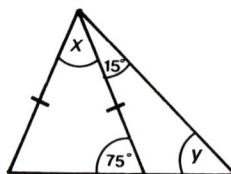

Fig. P47

9 A pipe has an internal radius of 8 cm and an external radius of 12 cm. It is 14 m long. Calculate the weight of the pipe in kg, correct to 3 s.f., if the metal from which it is made has a density of 7.5 g/cm³. (Take $\pi = \frac{22}{7}$.)

10 Draw four pairs of axes, each from −2 to 4. Find the equation of the straight line which passes through:
(a) (−1, −1) and (1, 3)   (b) (1, 1) and (2, 3)   (c) (1, 1) and (4, −2)
(d) (−2, 3) and (0, −1).

11 Solve graphically the simultaneous equations $y = x - 1$ and $y = 3x - 8$.

Paper Twenty-eight

1   Find the volume of a cylinder of height 42 cm and radius 8 cm, taking $\pi = \frac{22}{7}$.

2   What is the difference in volume between a 4 cm cube and 4 cm³?

3   In Figure P48, calculate:
   (a) BD   (b) ∠DBC
   (c) DC (Use your answers to parts (a) and (b).)
   (d) the area of the triangle.

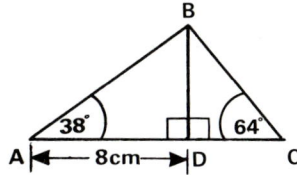

Fig. P48

4   Francesca needs an average of 50 marks over seven tests to gain a pass certificate. So far she has scored 32, 70, 41, 61 and 58. How many marks must she score in the last two tests if she is to just reach her target?

5   In marking an examination a teacher accepts that $\frac{15}{20} + \frac{29}{30} = \frac{44}{50}$. How can this be?

*6   A pyramid has a base 300 m square and a height of 150 m. Find its volume.

*7   Find the area of Figure P49, taking π as $\frac{22}{7}$.

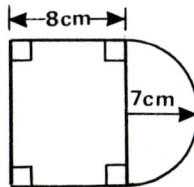

Fig. P49

8   How many revolutions does a cycle wheel of diameter 30 inches make in one mile? (36 inches = 1 yard; 1760 yards = 1 mile; take $\pi = 3\frac{1}{7}$.)

9   In Figure P50, TAM and TBN are tangents to the circle. Find the sizes of the lettered angles.

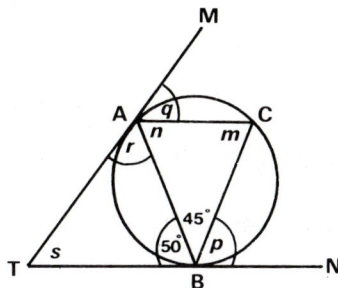

Fig. P50

10 State the equation of the line joining:

(a) $(-3, 3)$ to $(-3, -3)$ (b) $(-3, 3)$ to $(3, -3)$ (c) $(0, 4)$ to $(4, 0)$

(d) $(0, -2)$ to $(2, 4)$.

11

No. of questions wrong	0	1	2	3	4	5	6	7
No. of pupils (frequency)	5	7	11	15	14	10	3	1

(a) Find: (i) the mean   (ii) the median   (iii) the mode.

(b) Draw up a cumulative frequency table, then draw the cumulative frequency diagram.

Paper Twenty-nine

1 Referring to Figure P51:

(a) Write as a column matrix:

    (i) $\overrightarrow{AB}$ (ii) $\overrightarrow{BC}$ (iii) $\overrightarrow{AB} + \overrightarrow{BC}$

(b) Calculate $|\overrightarrow{AB}|$ (that is, the modulus, or magnitude, of $\overrightarrow{AB}$) correct to 2 decimal places.

(c) State $|\overrightarrow{AC}|$ correct to 2 decimal places.

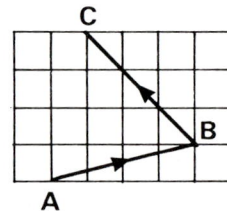

Fig. P51

2 Figure P52 shows a house extension.

(a) Find the area of the end wall, ABCD.

(b) Find the volume of the extension.

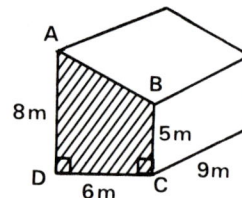

Fig. P52

3 (a) Write $3\,m^2$ in cm^2.   (b) Write $1\,cm^2$ in m^2.

*4 Find the value of n if:

(a) $2n = 3$ (b) $3n - 1 = 8$ (c) $\dfrac{4}{n} = 2$ (d) $\dfrac{6}{n} + 1 = 7$.

*5 Construct a trapezium with parallel sides of 4 cm and 3 cm, with the perpendicular distance between them being 2 cm. Then calculate its area.

*6 Taking $\pi = \frac{22}{7}$ find the total surface area of a cylinder 8 cm high with a radius of 10.5 cm.

7 State, as simply as possible, the equation solved by the x co-ordinates of the crossing points of $y = x^2 - x$ and $y = \frac{1}{2}x + 3$.

8 Find the values of x and y in Figure P53.

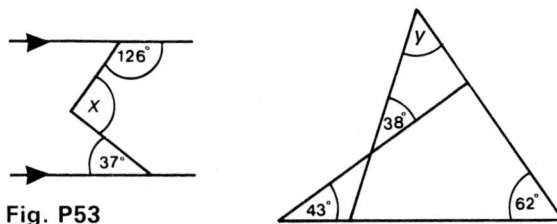

Fig. P53

9 Solve:
(a) $3(a - 1) + 4(2a - 1) = 26$ (b) $5(b - 4) + 28 = 3(b - 2)$.

10 Two ships set sail from the same port at 1000 hours. One sails NE at a speed of 12 knots and the other SE at 16 knots. Calculate how far apart they are at noon. (1 knot is 1 nautical mile per hour.)

11 Calculate the total compound interest on £280 invested at 8% for 4 years.

12 A cylindrical drum 24 cm high and of internal diameter 12 cm is half full of water.

(a) Calculate: (i) the capacity of the drum   (ii) the volume of the water.

(b) What is the volume of a sphere of radius 6 cm?

(c) If such a sphere is placed in the drum, show that the sphere and the water take up a total volume of about 2260 cm³, then calculate how far up the drum the water now reaches, assuming the sphere to be completely immersed in the water.

(Take $\pi = 3.142$. Give answers correct to 3 s.f.)

Paper Thirty

Note: These solids should be made in preparation for working Exercise 32A.

Use 1 cm-squared paper to copy the nets in Figure P54 (overleaf), marking them clearly with letters as shown. Each square drawn here represents a 1 cm square on your nets. Cut out the nets and make the solids. You (or your teacher) should keep them carefully for use when you work Exercise 32A.

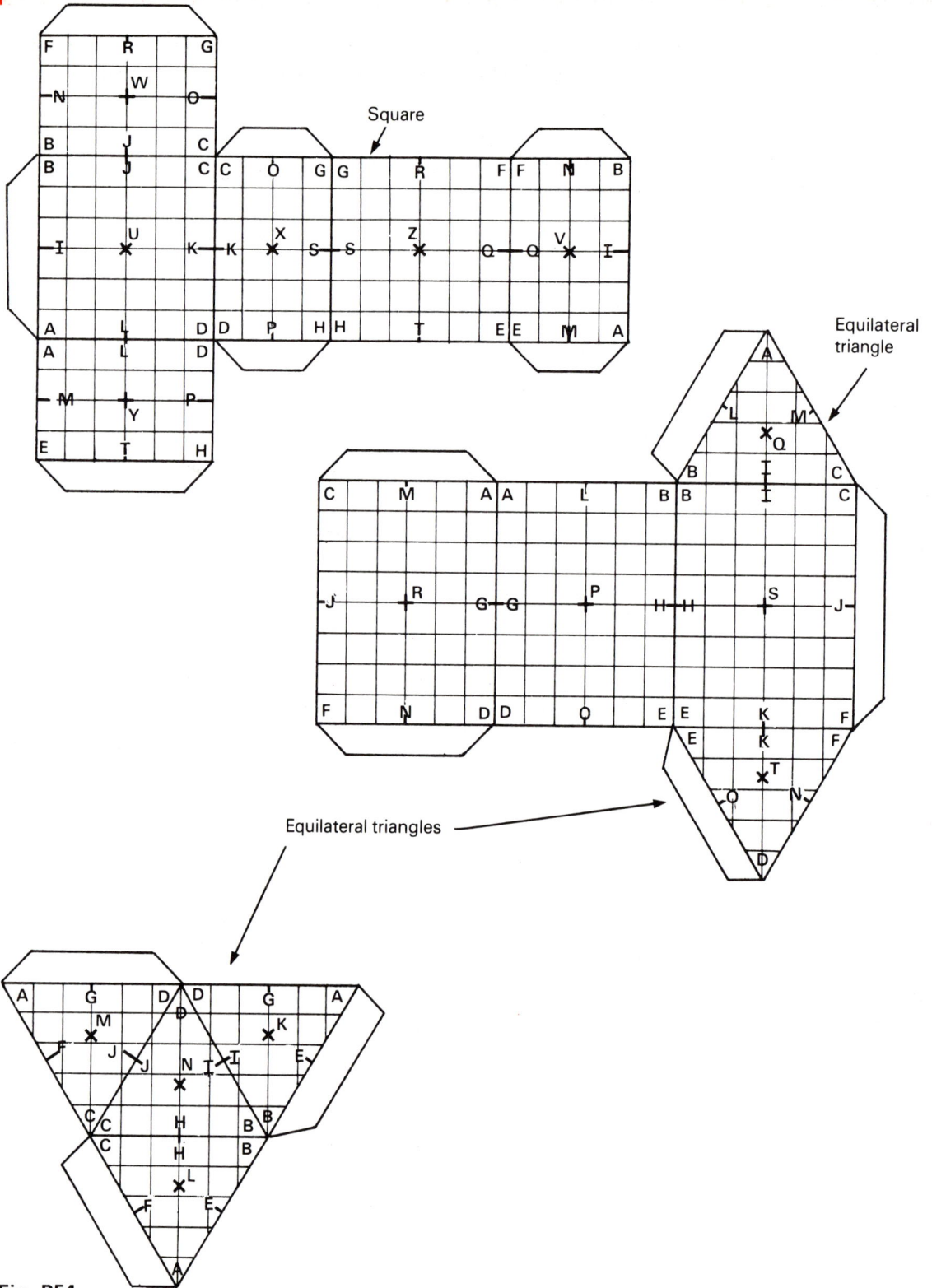

Square

Equilateral
triangle

Equilateral triangles

Fig. P54

Glossary

If you cannot remember what a word means, or cannot find a particular topic in the book, this glossary should help you. To save space, words that are listed alphabetically in the Index are not included here.

Notes: 'see 30', for example, refers you to Chapter 30;
'see 5A', for example, refers you to Exercise A in Chapter 5;
'see 5●', for example, refers you to '● You need to know' in Chapter 5.
Words in *italics* may be looked up in this glossary.

A

Adjacent	Next to: see 5●.
Allied	Joined to: see 5●.
Alternate	On opposite sides: see 5●.
Altitude	Height; a triangle has three altitudes, depending on which side is taken as the base line.
Amount	The whole, for example principal sum plus interest (see 30).
Apex	The top point, especially of a triangle.
Appreciation	Increase in value.
Associative	An *operation* (e.g. $+$, $-$, $\times$, $\div$) which is unaffected by brackets when three or more terms are combined, e.g. $+$ is associative as $3 + (4 + 5) = (3 + 4) + 5$, but $-$ is not as $3 - (4 - 5) \neq (3 - 4) - 5$.
Average	Usually the mean of a set of data, but mode and median are also averages in statistics (see 23●).

B

Bearing	The direction of one point from another. Cardinal bearings use N, E, S and W. Three-figure bearings are measured in degrees clockwise from north, e.g. east is 090° and west is 270°.
Binary	Base two; the number system that uses only the figures 0 and 1.

C

Coefficient	The *constant* multiplying an algebraic term, e.g. the 3 in $3x^2$.
Collinear	In the same straight line.
Column matrix	A matrix with only one column of figures, e.g. the vector $\begin{pmatrix} 2 \\ 3 \end{pmatrix}$.
Commission	Pay which depends on the value of the goods sold or services supplied.
Commutative	As *operation* (e.g. $+$, $-$, $\times$, $\div$) which is unaffected by the order of the terms: e.g. $+$ is commutative as $2 + 3 = 3 + 2$, but $-$ is not as $2 - 3 \neq 3 - 2$.

Complement	One of two parts that make up a whole. Complementary angles add up to 90°. Complement of a set: see 3●.
Concave	'With a cave'; a concave polygon has at least one angle pointing inwards ('re-entrant').
Concurrent	Meeting at the same point.
Congruent	Exactly the same both in shape and size: see 5A.
Conjecture	A guess or forecast without proof.
Consecutive	Following one after another: 5, 7, 9 are consecutive odd numbers. If n is even then n, $n + 2$ and $n + 4$ are consecutive even numbers.
Constant	Unchanging: a constant term in an algebraic expression will be the same for all values of the letters (the *variables*).
Construct	Draw accurately, usually using only a ruler and a pair of compasses.
Convex	'Pointing outwards'; the opposite of *concave*.
Corresponding	In the same position: see 5.
Cross-section	A cut across a solid.
Cubic	In the shape of, or involving, a cube. 1 cm³ (one cubic centimetre) is the volume of anything with the same volume as a cube of side 1 cm. A cubic expression involves a letter-term of *power* three, e.g. $4x^3 - 2x + 1$.
Cuboid	A solid with six rectangular faces, like a brick. If all the rectangles are squares it is called a cube.

D

Deduce	To reach a conclusion by reasoning.
Denary	Based on ten; our usual number system.
Denominator	The bottom number in a common fraction; it tells into how many parts the whole has been divided.
Deposit	An initial payment to reserve a piece of property: an amount paid into a savings account.
Depreciation	Loss in value.
Determinant	For matrix $\begin{pmatrix} a & b \\ c & d \end{pmatrix}$ the determinant is $ad - bc$. See 12A.
Diagonal	A line joining two corners of a polygon which are not *adjacent*.
Difference	The result of a subtraction.
Digit	One of the figures in a number.
Digit-sum	Used in this course for the result of continually adding the *digits* of a number until a single digit results: see 11●.
Discount	The amount by which a price is reduced.

Discrete	Able to be measured exactly, e.g. the number of pages in a book. The opposite is continuous, e.g. the length of a line, which can never be given exactly.
Dividend	The number being divided, e.g. the 8 in $8 \div 4$.
Divisor	The number you are dividing by, e.g. the 4 in $8 \div 4$.
Domain	The set of numbers, or the *region*, on which *operations* are performed.

E

Element	A member of a set; the things being combined in an *operation*.
Expand	In algebra, to multiply out the brackets, e.g. $(x + 3)(x - 3)$ expands to $x^2 - 9$.
Expression	A collection of terms with no equality, e.g. $4x - 2y^2 + 3z$.
Exterior angle	The angle between a *produced* side of a polygon and the *adjacent* side: see 22●.

F

Face	The flat side of a *polyhedron*.
Factors	Numbers or *expressions* that multiply by others to make the whole, e.g. 5 and 10 are factors of 100. For algebraic factors see 17, 25 and 29.
Fibonacci sequence	A sequence made by adding the previous two terms, usually starting 1, 1, 2, 3, 5, 8.
Formulate	To express in a clear or definite form.
Frustum	The part of a pyramid or cone contained between the base and a plane parallel to the base.
Function	An algebraic 'event', e.g. a function of x is such that any value, x, becomes x^2, often written as $f(x) : x \rightarrow x^2$, or $f(x) = x^2$. Then $f(2) = 4$, as 4 is 2^2.

G

Generalise	Express in general terms, usually using an algebraic formula.
Gradient	The slope of a line; in mathematics the tangent of the angle made with the horizontal. See under $y = mx + c$.
Gross	The whole amount, before any deductions.

H

Hatch	To define an area by drawing a set of parallel sloping lines on it.
Hexadecimal	Base sixteen, which uses the letters A to F to represent 10 to 15 in machine-code programming of computers.
Histogram	A bar-chart in which the area of the bars represents the frequency. There is no vertical scale, but a key showing the area that represents one unit.
Hypotenuse	The longest side in a right-angled triangle: see 8●.

I

Image	The result of a transformation: see 1.
Improper fraction	A top-heavy fraction, like $\frac{13}{2}$.
Inclusive	'Including everything', e.g. {integers from 1 to 3 inclusive} = {1, 2, 3}.
Infinite	Without ending.
Intercept	Part of a line between two crossing points.
Interior angle	The angle made inside a polygon by two *adjacent* sides: see 22●.
Intersection	The crossing point; see also 3●.
Invoice	A bill, setting out the payment required.
Irrational	Unable to be written as a common fraction, e.g. π, $\sqrt{2}$.

L

Litre	The metric unit for capacity; 1 litre = 1000 cm³ (or c.c.); 1 ml = 1 cm³.
Locus	The path made by a moving point.
Loss/Loss%	The opposite of *profit*. A car bought for £8000 and sold for £2000 is sold at a loss of £6000 or $\frac{6000}{8000} \times 100\% = 75\%$.

M

Mixed number	A number consisting of an integer and a fraction, like $3\frac{3}{4}$.
Modulus	See 28●.
Multiple	A number made by multiplying one integer by another.
Multiplying factor	The fraction used to increase or decrease in a given ratio, e.g. to change in the ratio $x:y$ use the multiplying factor $\frac{x}{y}$.

N

Natural numbers	Numbers used for counting.
Net	The plane shape that folds to make a solid. The amount left after deductions.
Numerator	The top number in a fraction.

O

Octal	Base eight, using the figures 0 to 7 only.
Operation	A way of combining elements. It may be a standard method, such as +, −, × or ÷, or a combination defined by a special sign, often * or ∘. For example $a * b$ could be defined to mean double a then add the square of b, so that $2 * 4 = 20$.
Ordered pair	An alternative name for co-ordinates.

P

Percentage error	An indication of the seriousness of an arithmetical error. Found by the formula: (Error/True) × 100%.
Perpendicular	At right angles.
Perpendicular bisector	A line which crosses the midpoint of another line at right angles: see 5●.
Pictogram	A chart showing information by means of picture symbols that represent a certain amount.
Point symmetry	A figure has point symmetry if it looks the same after a rotation of 180°.
Polyhedron	A solid with flat faces; the five regular ('Platonic') polyhedra are: tetrahedron (4 equilateral triangles); cube (6 squares); octahedron (8 equilateral triangles); dodecahedron (12 regular pentagons); icosahedron (20 equilateral triangles).
Power	The result of multiplying a number by itself a number of times. It can be shown using an index (raised figure), e.g. as 5^3 (the third power of 5).
Premium	Payment made for insurance.
Principal	Money on which interest is paid: see 30.
Prism	A solid with a constant *cross-section.*
Produce	To make a line longer, e.g. 'Produce AB to C' means lengthen line AB from end B until it reaches point C.
Product	The result of multiplying numbers.
Profit	The gain made when something is sold for more than was paid for it. Percentage profit is found from the formula: (Profit/Cost price) × 100%.
Proportion	Two quantities are in proportion when corresponding pairs are in the same ratio, e.g. m.p.h. and km/h. They give a straight-line conversion graph. See 9●.
Pyramid	A solid with a polygon as its base and triangular sides meeting at a common *vertex.*

Q

Quadratic equation	An equation involving a squared term, e.g. $3x^2 - 3 = 4x$.
Quotient	The result of a division.

R

Rank	To place in order, e.g. of size.
Rateable value	The amount allotted to a property on which the rates to be paid to the council depend. For example, if a house has a rateable value of £300 and the rate is 50p in the £, then £150 will have to be paid.
Rational	Able to be written exactly as a common fraction.

Real number	A number that exists and can be written as a (possibly endless) decimal fraction; π is real but $\sqrt{-1}$ is not.
Reciprocal	One over a number, e.g. the reciprocal of 2 is $\frac{1}{2}$ or 0.5.
Rectangular numbers	Numbers that are not prime (except 1).
Recur	To repeat, as in recurring decimals, the repeating figure(s) being indicated with a dot (or two dots).
Reflex	An angle more than 180°. If a reflex angle is intended then it will be written as 'Angle ABC reflex'.
Region	A special area of a diagram, especially of a graph or network. See 16●.
Rotational symmetry	The number of times a shape fits into a tracing of itself in one full rotation is the order of rotational symmetry, e.g. a square is of order 4. Rotational symmetry of order 1 is often not counted as true rotational symmetry.
Round off	To approximate.

S

Scientific notation	See 6.
Segment	Part of a circle cut off by a chord: see 18A.
Similar	Exactly the same shape. Two similar objects have *corresponding* sides in the same ratio. For similar solids with sides' ratio $x:y$, their areas are in the ratio $x^2:y^2$ and their volumes are in the ratio $x^3:y^3$.
Simplify	In algebra, to make an *expression* simpler, usually by multiplying out brackets and collecting like terms.
Slant height	The shortest distance from the *apex* of a *pyramid* to the base, measured along a *face*.
Solve	To find the numerical value of the letter(s).
Sphere	The mathematical name for a ball.
Subtend	Angle APB is subtended at point P by the lines AP and BP. See 13B.
Sum	The result of an addition.
Supplementary	Adding up to 180°.
Surd	An *irrational* number.

T

Tangent	In geometry, a straight line which touches, but does not cross, a curve. A trigonometrical ratio: see 8.
Taxable income	The amount of your pay on which you have to pay tax; you are allowed to earn a certain amount before you have to pay any income tax.
Tonne	A metric unit of weight (mass) equal to 1000 kg.

Translate	To slide a shape: see 28●.
Triangular number	One of a sequence of numbers of dots that make equilateral triangles. The sequence starts 1, 3, 6, 10, 15 . . . The nth term is $\frac{1}{2}n(n + 1)$.

V

Variable	A letter which may stand for various numbers.
Vertex	A corner of a shape.
Vulgar fraction	A common fraction that is not top-heavy.

Y

$y = mx + c$	The general equation of all straight-line graphs. m is the slope or *gradient* and c is the crossing point on the y-axis. See 16●.

Answers

Note: Answers not printed here are given in the Teacher's Manual.

Chapter 1

Test yourself

1. A, (2, 2); B, (1, 0); C, (−2, 2); D, (0, −2)
2. (a) B (b) D (c) A (d) C
3. (4, 4); (2½, 2½); (0, 0); (−3, −3); (−1, −1)
4. (4, −4); (−2½, 2½); (0, 0); (−3, 3); (1, −1)
5. (a) 14 (b) (4 7) (c) $\begin{pmatrix} 2 \\ 3 \end{pmatrix}$

    (d) $\begin{pmatrix} 0 & -3 & 1 \\ 0 & 0 & 0 \end{pmatrix}$ (e) $\begin{pmatrix} 1 & -1 & 0 \\ -1 & 1 & 2 \end{pmatrix}$

6. (a) $\begin{pmatrix} -1 & 1 & -1 \\ 2 & 0 & 1 \end{pmatrix}$

    (b) There must be the same number of columns in the first matrix as rows in the second.
7. (a) Rotation of 180° about (0, 0).
    (b) Reflection in $y = -x$.
    (c) Enlargement, scale factor −1, centre (0, 0).
    (d) Reflection in $y = 0$ followed by reflection in $x = 0$; reflection in $x = 0$ followed by reflection in $y = 0$.
8. See Figure A1:1.

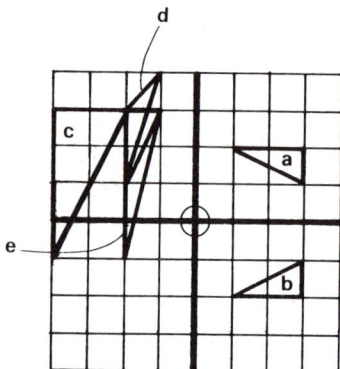

Fig. A1:1

Exercise 1A

6. (a) Shear, invariant line $x = 0$; (1, 0) → (1, −1).
    (b) Shear, invariant line $x = 0$; (1, 0) → (1, −2).
    (c) Shear, invariant line $y = 0$; (0, 1) → (3, 1).
    (d) Shear, invariant line $y = 0$; (0, 1) → (−1, 1).
    (e) Shear, invariant line $y = 0$; (0, 1) → (2, 1).
    (f) Shear, invariant line $y = 0$; (0, 1) → (−2, 1).
    (g) Enlargement, scale factor −2; centre (0, 0).
    (h) One-way stretch from $x = 0$, stretch factor 2.
    (i) One-way stretch from $y = 0$, stretch factor 2.

Exercise 1B

5. (a) One-way stretch from $x = 0$, stretch factor 2.
    (b) One-way stretch from $x = 0$, stretch factor 3.
    (c) One-way stretch from $y = 0$, stretch factor 2.
    (d) One-way stretch from $y = 0$, stretch factor 3.
    (e) Two-way stretch from $x = 0$, stretch factor 2 and from $y = 0$, stretch factor 3.
    (f) Two-way stretch from $x = 0$, stretch factor 3 and from $y = 0$, stretch factor 2.

Chapter 2

Test yourself

1. (a) −3 (b) 5 (c) 1 (d) 2 (e) −3
    (f) −3 (g) 2 (h) −3 (i) $\frac{1}{2}$
2. (a) $3x + 12$ (b) $2a − 10$ (c) $12 + 8a$
    (d) $4a − 10$ (e) $−3 − a$ (f) $−h + 1$
    (g) $−6x − 3$ (h) $−2x + 8$ (i) $−9 − 12a$
3. (a) −3 (b) $\frac{1}{4}$ (c) 3 (d) 1 (e) −2
    (f) −7 (g) −6 (h) $1\frac{1}{4}$ (i) $−2\frac{1}{2}$ (j) $5\frac{1}{3}$
    (k) −3 (l) $−\frac{2}{3}$
4. (a) $6 + 9a$ (b) $9a − 12$ (c) $−x − 1$
    (d) $−2x − 8$ (e) $−8 + 4a$ (f) $−4a + 6$
    (g) $6a − 16$
5. (a) 4 (b) 2 (c) −7 (d) −2 (e) −2
    (f) 3 (g) $8\frac{1}{2}$ (h) −1 (i) $−2\frac{1}{3}$ (j) −3
    (k) −4 (l) 2
6. (a) 12 (b) −16 (c) 5 (d) −4 (e) 1
7. (a) 32 (b) 10 (c) 12 (d) 3 (e) 2 (f) $2\frac{1}{2}$
8. (a) $x = 9, y = 6$ (b) $x = 2, y = 2$
    (c) $x = 3, y = 8$
9. (a) $x = 1, y = -1$ (b) $x = 4, y = 1$
    (c) $a = 9\frac{1}{2}, b = 5\frac{1}{2}$ (d) $s = \frac{5}{11}, t = -1\frac{9}{11}$
10. (a) $12 − a$ (b) $7a + 18$ (c) $a − 35$
    (d) $26 − 4a$ (e) $36a − 25$ (f) $4 − 18a$
11. (a) $−6\frac{2}{3}$ (b) $\frac{1}{2}$ (c) $3\frac{2}{7}$ (d) $\frac{2}{63}$
12. (a) $a = 3, b = 6$ (b) $a = 10, b = 1$
    (c) $a = 2, b = 1\frac{1}{2}$ (d) $a = 12, b = -5$

Exercise

3. (a) 10 (b) 1 (c) 4
4. (a) −1 (b) −1 (c) 6

5 (a) 9 (b) 2
6 (a) 6 (b) 9 (c) 0
7 (a) 5 (b) $7\frac{1}{2}$ (c) 0 (d) 1.2
8 (b) Dad is twice as old as Josie.
(c) When Josie is 10.
(d) $1\frac{1}{2}(x + 5) = 25 + x \Rightarrow x = 35$; when Josie is 40 and her Dad is 60.
9 74 min
10 4 units per side
11 $x = 6$; sides 4 cm, 4 cm, 11 cm, which is not a triangle!
$x = 3$; sides 2 cm, 2 cm, 1 cm
$x = 2\frac{1}{2}$; sides $\frac{1}{2}$ cm, $\frac{1}{2}$ cm, $1\frac{2}{3}$ cm, which is not a triangle!
12 $x = 20$ cm; $y = 10$ cm.

Chapter 3

Test yourself

1 (a) T (b) F (c) F (d) T (e) T (f) F
(g) T (h) T (i) T (j) F
2 See Figure A3:1.

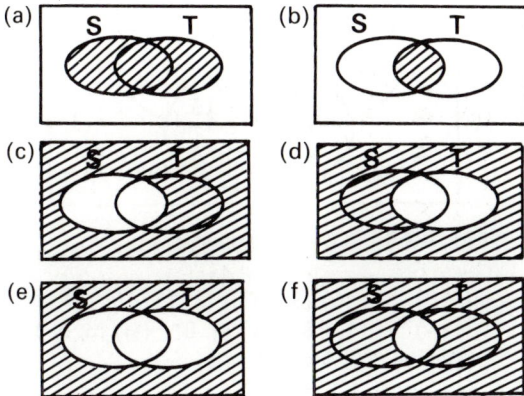

Fig. A3:1

3 See Figure A3:2.

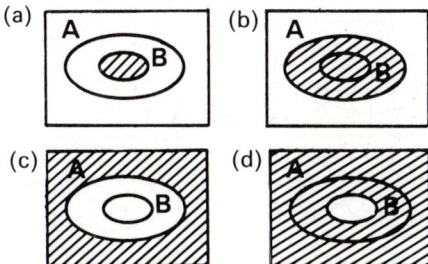

Fig. A3:2

4 (a) $\varnothing$ (b) {1, 2, 5} (c) {5, 8} (d) {1, 2, 8}
(e) {1, 2, 5, 8} (f) {8}
5 (a) {5} (b) {2, 3, 5} (c) {3, 4} (d) {2, 4}
(e) {2, 3, 4} (f) {4}
6 See Figure A3:3.

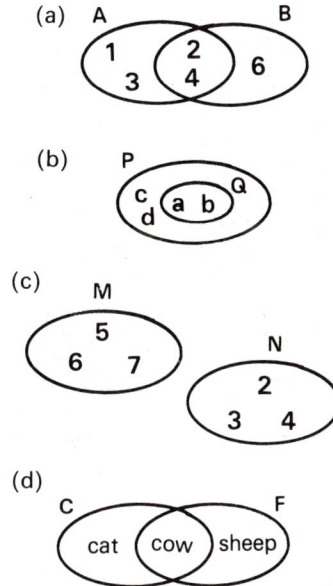

Fig. A3:3

7 (a) 20 (b) 11 (c) 12 (d) 7 (e) 16 (f) 9
(g) 8 (h) 13 (i) 4
8 (a) W (b) Z (c) X (d) V (e) Y
9 See Figure A3:4.

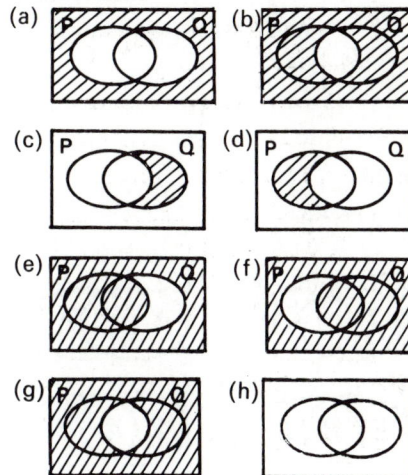

Fig. A3:4

10 See Figure A3:5.

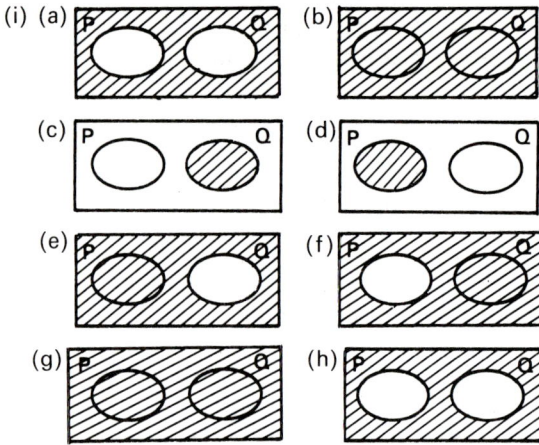

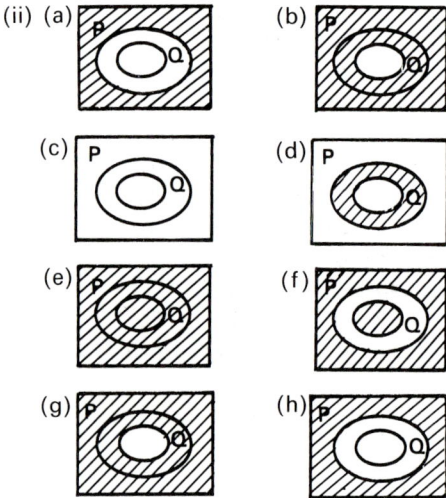

(i) (a) (b)

(c) (d)

(e) (f)

(g) (h)

(ii) (a) (b)

(c) (d)

(e) (f)

(g) (h)

Fig. A3:5

Exercise 3A

1 See Figure A3:6.

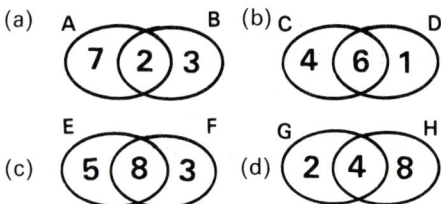

(a) (b)

(c) (d)

Fig. A3:6

2 (a) 4 (b) 4
3 (a) 6 (b) 8
4 48
5 (a) 9 (b) 6 (c) 3 (d) 12

Exercise 3B

4 (a) See Figure A3:7.

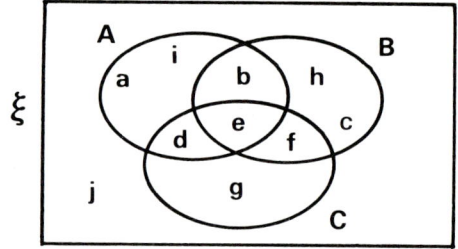

Fig. A3:7

(b) (i) {e} (ii) {b, e} (iii) {d, e}
(iv) {e, f} (v) {j}

5 (a) See Figure A3:8.

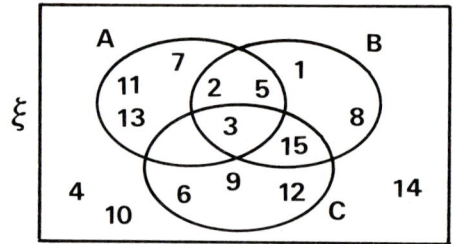

Fig. A3:8

(b) (i) {3} (ii) {3} (iii) {3, 15}
(iv) {7, 11, 13} (v) {1, 8} (vi) {6, 9, 12}
(vii) {4, 10, 14}

6 See Figure A3:9.

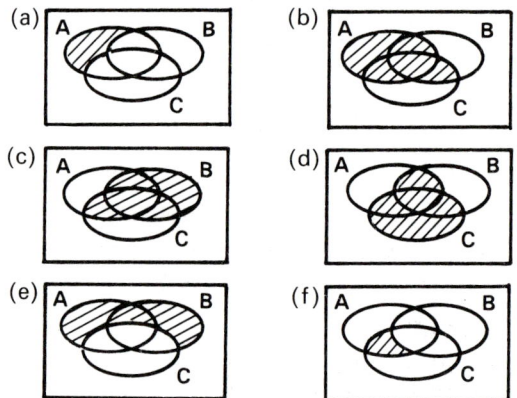

(a) (b)

(c) (d)

(e) (f)

Fig. A3:9

Exercise 3C

5 21
6 (a) 5 (b) 7 (c) 11 (d) 13
7 (a) 1 (b) 29
8 (a) 8 (b) 2
9 (a) 9 (b) 20
10 (a) 1 (b) 15
11 (a) 6 (b) 2

Chapter 4

Test yourself

1 (a) 2.923 65 (b) 0.799 043 (c) −0.2166
2 (a) 1.357 224 (b) 36.430 69 (c) 60.030 77
  (d) 25.124 75 (e) 6.790 0915 (f) 4.932 72
3 (a) 0.115 (b) 0.738 (c) 54.8 (d) 77.2
  (e) 22.3 (f) 7.04 (g) 1.32 (h) 20.3
  (i) 0.153

Exercise

1 (a) 19 000 (b) 30 400 (c) 1.12
2 (a) 2.71 (b) 3.50 (c) 5.67 (d) 2.13
  (e) 2.60 (f) 6.69
3 (a) 0.641 (b) 0.138 (c) 3 (d) 0.667
  (e) 0.571
4 (a) (i) 8 h 36 min 52 s (ii) 11 h 59 min 21 s
    (iii) 27 h 51 min 41 s (iv) 24 h 45 min 35 s
    (v) 8 h 20 min 16 s (vi) 58 h 23 min 13 s
  (b) (i) 46° 48′ (ii) 60° 15′ (iii) 41° 52′
    (iv) 163° 46′
  (c) (i) 32.3° (ii) 21.1° (iii) 80.7° (iv) 18.9°
  (d) (i) 169 yds 1 ft 6 ins
    (ii) 225 yds 2 ft 10 ins
    (iii) 277 yds 2 ft 4 ins

Chapter 5

Exercise 5A

3 (a) Not congruent (angle not included).
  (b) $\frac{UVW}{XWV}$; right angle, hypotenuse, side.
  (c) Not congruent (no side given).
  (d) $\frac{DEF}{HGF}$; 2 sides, included angle.
  (e) Not congruent (side not corresponding).
  (f) $\frac{MNP}{ONP}$; 2 sides, included angle.
4 (c) (i) $\angle ABE = \angle FBC = 90° + \angle CBE$
5 (b) $BD = DC$; $\angle BAD = \angle CAD$

Exercise 5B

3 (a) $AB = 1\frac{1}{2}$ cm, $BZ = 2\frac{1}{4}$ cm
  (b) $HI = 9\frac{1}{3}$ cm, $EG = 6\frac{3}{4}$ cm
  (c) $LN = 6\frac{3}{4}$ cm, $LM = 4\frac{1}{2}$ cm

4 Alternate angles are only equal if between
  parallel lines.
5 $ED = 15$ cm, $BC = 3.6$ cm
6 $ST = 3\frac{3}{4}$ cm, $TV = 6$ cm
8 (a) $AY = 3\frac{1}{2}$ cm, $XY = 4$ cm

Chapter 6

Test yourself

1 (a) 220 (b) 213
2 £10; £9.97
4 (a) 17.14 (b) 0.78 (c) 1.01 (d) 10.11
  (e) 19.11 (f) 206.80
5 (a) 17 (b) 0.78 (c) 1.0 (d) 10
  (e) 19 (f) 210

Exercise 6A

4

Sun	–	1.4×10^6
Mercury	5.8×10^7	4.9×10^3
Venus	1.1×10^8	1.2×10^4
Earth	1.5×10^8	1.3×10^4
Mars	2.3×10^8	6.8×10^3
Jupiter	7.8×10^8	1.4×10^5
Saturn	1.4×10^9	1.2×10^5
Uranus	2.9×10^9	5.2×10^4
Neptune	4.5×10^9	5.0×10^4
Pluto	5.9×10^9	6.0×10^3

5 4×10^0 ($10^0 = 1$)
6 (a) 2.45×10^5 (b) 2.28×10^6
  (c) 1.78×10^7 (d) 6.51×10^{10}
  (e) 3.296×10^8 (f) 8.82×10^{14}
7 (a) 950 000; 1 050 000
  (b) 995 000; 1 005 000
  (c) 999 500; 1 000 500
  (d) 999 950; 1 000 050
  (e) 999 995; 1 000 005

Exercise 6B

3 Energy 1.678×10^3
  Protein 1.22×10
  Fat 8.6×10^0
  Carbohydrate 7.2×10
  Vitamins: A, 1.2×10^{-3} B1, 1.8×10^{-3}
        B2, 2.3×10^{-3} C, 4.2×10^{-2}
        D, 1.7×10^{-5}
  Niacin 2.7×10^{-2} Iron 2×10^{-2}
  Calcium 1.2×10^0
4 (a) 7.9×10^2 (b) 8.0×10^3
  (c) 4×10^2 (d) 2.50×10
  (e) 1.04×10^2 (f) 6.1×10^4
  (g) 3.17×10^3 (h) 4.3×10^3
  (i) 4.00×10^6
5 (a) 7.6×10^{-1} (b) 9.9×10^{-3}

6 (a) 8.2×10^{-1} (b) 1.64×10
(c) 1.44×10 (d) 1.23×10^{-3}
(e) 2×10^5 (f) 4×10^0
(g) 2.5×10^5 (h) 2×10^{-1}
(i) 3.1×10 (j) $3.\dot{3} \times 10^{-5}$

7 (a) $\frac{1}{3}$ (b) $\frac{1}{9}$ (c) $\frac{1}{16}$ (d) $\frac{1}{8}$ (e) $\frac{1}{81}$ (f) $\frac{1}{400}$

Chapter 7

Test yourself

1 (a) $t = a + g$ (b) $t = h - s$
(c) $t = \dfrac{f + y}{2}$ (d) $t = \dfrac{4w - r}{3}$
(e) $t = \dfrac{w - ag}{e}$ (f) $t = \dfrac{4k - a}{3}$

2 (a) $c = 2(g - f) - a$ (b) $c = \sqrt{h - a^2}$
(c) $c = \left(\dfrac{t}{2\pi}\right)^2$ (d) $c = \frac{1}{2}\left(R - \dfrac{d}{ab}\right) = \dfrac{abR - d}{2ab}$
(e) $c = \dfrac{s}{r(a - t)}$
(f) $c = \dfrac{3d - t}{f} - b = \dfrac{3d - t - bf}{f}$

3 (a) $r = \dfrac{3a + 2gp}{2g + 1}$
(b) $r = \dfrac{-st}{2 + bc - a} = \dfrac{st}{a - bc - 2}$
(c) $r = \dfrac{at}{a + t}$ (d) $r = \dfrac{bs}{2 - 3as}$
(e) $r = \dfrac{18w^2}{s^2}$

Exercise 7A

4 40 cm
5 $\theta = \dfrac{360 \times A}{C}$; $120°$
6 (a) (i) $D = \dfrac{360 \times A}{\theta \times \pi}$ (ii) $\theta = \dfrac{360 \times A}{\pi \times D}$
(b) 9 cm (c) 90°
7 (a) 57.3° (b) 24 cm

Exercise 7B

2 54.4 cm; 172 cm^2
4 10 cm
5 (b) 11.25 cm^2 (c) 6.25^c
6 126.6 cm^2

Chapter 8

Test yourself

1 (a) 6.7 cm (b) 6.4 cm (c) 5.7 cm
(d) 5.2 cm (e) 72 m (f) 5.6 cm
(g) 13.0 cm (h) 11 m

Exercise 8A

4 (a) 40.3 cm; 48.5 cm (b) 15.9 cm; 25.4 cm
(c) 20.5 cm; 24.6 cm
5 (a) 0.9083 (b) 1.387 (c) 0.3395
(d) 0.4752 (e) 5.769
6 (a) 8.39 (b) 1.60 (c) 6.20
(d) 14.1 (e) 11.4
7 1.1 metres
8 (a) 5.6 n.m. (b) 9.8 n.m.

Exercise 8B

3 (a) 56.3° (b) 8.5° (c) 66.0°
4 (a) $\angle$P = 33.3°; $\angle$Q = 56.7°
(b) $\angle$P = 39.8°; $\angle$Q = 50.2°
(c) $\angle$P = 25.3°; $\angle$Q = 64.7°
(d) $\angle$P = 70.0°; $\angle$Q = 20.0°

Chapter 9

Test yourself

1 (a) (i) 7:10 (ii) $1:1\frac{3}{7}$ (iii) $\frac{7}{10}:1$
(b) (i) 25:1 (ii) $1:\frac{1}{25}$ (iii) 25:1
(c) (i) 200:3 (ii) $1:\frac{3}{200}$ (iii) $66\frac{2}{3}:1$
(d) (i) 1:600 (ii) 1:600 (iii) $\frac{1}{600}:1$
2 £21.60
3 320 g
4 £150 and £210
5 640 g nettle tops   50 g butter
2 medium onions $1\frac{1}{4}$ litres stock
25 cl cream
Leaves of spinach, salt, pepper, lemon juice to taste
Cook for 30 min
Serves 8
6 (a) 2:1 (b) 3 g
7 £4.63
8 (a) Army, 900 000 dollars; Air Force, 1 500 000 dollars; Navy, 600 000 dollars
(b) 650 000 dollars
9 £1260
12 9:12:14

Exercise 9A

13 96 cm^2
14 (a) 48 cm^2 (b) 108 cm^2 (c) 3 cm^2
15 4 cm
16 (a) 30 m (b) 500 m (c) 1.8 km (d) 7 km
(e) 175 m (f) 2.3 km
17 (a) 40 km by 40 km (b) 1 km and 10 km
(c) (i) 1600 km^2 (ii) 1 km^2 (iii) 100 km^2
18 (a) 0.25 km^2 (b) 1.84 km^2

Exercise 9B

8 (a) 240 g (b) 10 mm radius

9 (a) 125 ml (b) 3 cm
10 (a) (i) 2 : 1 (ii) 8 cm³
  (b) (i) 2 : 1 (ii) 256 cm³
  (c) (i) 5 : 4 (ii) 10.24 cm³
  (d) (i) 5 : 7 (ii) 35.7 cm³
11 About 10 kg and about 2 kg.
12 840 calories; 2000 calories
13 13.5 cm; $2\frac{1}{4}$ times as much card.
14 About 100 cm by 50 cm.
15 (a) 3 cm (b) About 2.8 p.
16 (a) 16.3 cm (b) 20.5 cm
17 (a) 163 kg (b) 27 stone
  Clearly adults and babies are not similar shapes.
18 (a) (i) 7 : 3 (ii) 49 : 9 (iii) 343 : 27
  (b) $2\frac{1}{3}$
  (c) $2286\frac{2}{3}$ pounds (about a tonne)

Chapter 10

Test yourself

1 (a) $\frac{1}{4}$ (b) $\frac{1}{16}$ (c) $\frac{1}{64}$
2 (a) $\frac{1}{2}$ (b) $\frac{1}{3}$ (c) $\frac{1}{6}$ (d) $\frac{1}{4}$ (e) $\frac{1}{9}$ (f) $\frac{1}{12}$
3 (a) $\frac{1}{4}$ (b) $\frac{1}{16}$ (c) $\frac{1}{64}$ (d) $\frac{1}{13}$ (e) $\frac{1}{169}$ (f) $\frac{1}{2704}$
  (g) $\frac{3}{13}$ (h) $\frac{1}{52}$ (i) $\frac{1}{16}$ (j) $\frac{1}{104}$ (k) $\frac{1}{676}$
4 (c) $\frac{2}{11}$ (d) $\frac{5}{22}$ (e) $\frac{1}{11}$ (f) $\frac{1}{11}$
5 (b) $\frac{1}{17}$ (c) $\frac{11}{850}$ (d) $\frac{1}{13}$ (e) $\frac{1}{221}$ (f) 0
6 Final two columns:

RBR	$\frac{1}{55}$	BRG	$\frac{1}{22}$	GRB	$\frac{1}{22}$
RBB	$\frac{3}{110}$	BBR	$\frac{3}{110}$	GRG	$\frac{1}{22}$
RBG	$\frac{1}{22}$	BBB	$\frac{1}{55}$	GBR	$\frac{1}{22}$
RGR	$\frac{1}{44}$	BBG	$\frac{1}{22}$	GBB	$\frac{1}{22}$
RGB	$\frac{1}{22}$	BGR	$\frac{1}{22}$	GBG	$\frac{2}{33}$
RGG	$\frac{1}{22}$	BGB	$\frac{1}{22}$	GGR	$\frac{1}{22}$
BRR	$\frac{1}{55}$	BGG	$\frac{2}{33}$	GGB	$\frac{2}{33}$
BRB	$\frac{3}{110}$	GRR	$\frac{1}{44}$	GGG	$\frac{1}{22}$

Exercise

7 (a) $\frac{1}{12}$ (b) $\frac{1}{9}$ (c) $\frac{3}{4}$ (d) $\frac{7}{8}$
8 (a) $\frac{5}{18}$ (b) $\frac{25}{72}$
9 (a) $\frac{1}{6}$ (b) $\frac{5}{24}$ (c) $\frac{1}{9}$ (d) $\frac{5}{24}$ (e) $\frac{2}{9}$ (f) $\frac{9}{16}$
10 (a) $\frac{1}{26}$ (b) $\frac{1}{26}$ (c) $\frac{1}{52}$ (d) $\frac{1}{13}$ (e) $\frac{1}{13}$ (f) $\frac{1}{26}$
11 (a) (i) $\frac{1}{6}$ (ii) $\frac{1}{3}$ (b) (i) $\frac{1}{8}$ (ii) $\frac{1}{6}$ (c) $\frac{5}{6}$
12 Fifteen ways; $\frac{5}{72}$

Chapter 11

Test yourself

15 (a) −8 (b) −30 (c) 12 (d) −3 (e) −2
  (f) 21 (g) 11 (h) −9
17 (a) 3; 30 (b) 4; 480

18 (a) {7.9, 690, 7100}
  (b) {7.88, 691.40, 7108.80}
19 (a) £7.04 (b) £5.72 (c) £9.02
20 (a) 3.5 cg (b) 78 mm (c) 0.350 km
  (d) 30 litres (e) 5000 kg (f) 4650 m
  (g) 7.5 mm (h) 19 cm (i) 50 000 cm
  (j) 1000 cm³ (k) 10 000 cm²
  (l) 1 000 000 cm³
21 £364.80
22 576 days
23 £930.80
24 £8.14
25 (a) 3.6×10^8 km (b) about $3\frac{1}{2}$ h
26 (a) 2 kg (b) 1.5 litres
27 £112.50
28 1.5×10^{10} litres
29 7.65 cm; 2.85 cm; 21.8025 cm²
30 (a) 400 (b) 74.4 metres

Chapter 12

Exercise 12A

3 (a) $\frac{1}{2}\begin{pmatrix} -2 & 4 \\ -2 & 3 \end{pmatrix}$ (b) $\frac{1}{6}\begin{pmatrix} -2 & 4 \\ -3 & 3 \end{pmatrix}$
  (c) no inverse (d) $\frac{1}{7}\begin{pmatrix} 1 & 5 \\ -1 & 2 \end{pmatrix}$
  (e) $\frac{1}{3}\begin{pmatrix} 1 & 3 \\ -3 & -6 \end{pmatrix}$ (f) $\frac{1}{2}\begin{pmatrix} -3 & 2 \\ -7 & 4 \end{pmatrix}$
  (g) no inverse (h) $\frac{1}{2}\begin{pmatrix} -3 & 2 \\ 5 & -4 \end{pmatrix}$
  (i) $\frac{1}{17}\begin{pmatrix} -2 & -5 \\ 1 & -6 \end{pmatrix}$

Exercise 12B

1 (a) $x = 2, y = 5$ (b) $x = 1, y = 2$
2 (a) $x = 3, y = 2$ (b) $x = 4, y = 2$
  (c) $x = 5, y = 2$ (d) $x = 2, y = 5$
  (e) $x = 1.5, y = -2$ (f) $x = -2.5, y = -3.5$

Chapter 13

Exercise 13A

5 (a) 4.33 cm (b) 9.46 cm (c) 15.1 cm
6 (a) a circle (b) 10 cm − 8 cm = 2 cm
  (c) 10 cm + 8 cm = 18 cm
7 4 cm + 3 cm = 7 cm; 4 cm − 3 cm = 1 cm

Exercise 13B

4 15°
5 45°; 60°; 75°

Exercise 13C

3 (a)(i) 90° (∠ in semicircle)
  (ii) 50° (∠ sum of triangle)
  (iii) 50° (Alt. ∠s, CB // AD)
  (vi) 50° (∠s of isos. Δ)
  (v) 80° (∠ sum of Δ)
  (b) ∠CAD = 40° + 50° = 90°
  ∴ CADO is a semicircle.
  (c) Reasons as (a): (i) 90° (ii) 55° (iii) 55°
  (iv) 55° (v) 70°

4 5.7 cm

Chapter 14

Exercise

11 (a) 4.20 cm (b) 48°11′
12 (a) 19.2° (b) 3.29 cm (c) 3.29 cm
  (d) 9.44 cm (e) 70.8° (f) 70.8° (g) 47.6°
  (h) 47.6° (i) 42.4° (j) 42.4°
13 (a) 69.5° (b) 20.5° (c) 69.5° (d) 4.20 m
  (e) 11.2 m (f) 1.57 m (g) 4.48 m
14 FH = 12 m; EG = 13 m
  ∠FHD = ∠EDF = ∠GFH = 22°37′
  ∠FDH = ∠EFD = ∠FHG = 67°23′
  EF = 1.92 m; FG = 11.08 m
  ED = GH = 4.62 m

15 (a) (i) $\theta + \alpha = 90°$ (ii) $\sin \theta = \dfrac{AB}{AC}$

  (iii) $\cos \theta = \dfrac{AB}{AC}$

  (b) (i) 30° (ii) 40° (iii) 16°42′ (iv) 9°55′

Chapter 15

Exercise 15A

4 (a) $b^2 + b - 20$ (b) $a^2 + 10a + 16$
  (c) $a^2 - 8a + 15$ (d) $a^2 - 4a - 32$
5 (a) $r^2 - 3r - 28$ (b) $p^2 + 12p + 36$
  (c) $b^2 - 6b + 9$ (d) $m^2 - 14m + 49$
6 (a) $15 + 8x + x^2$ (b) $8 + 2x - x^2$
  (c) $15 - 2x - x^2$ (d) $14 + 5x - x^2$
  (e) $6 - 5x + x^2$ (f) $16 - 8x + x^2$
7 (a) $x^2 + ax + bx + ab$
  (b) $x^2 - ax + bx - ab$ (c) $x^2 + ax - bx - ab$
  (d) $x^2 - ax - bx + ab$
8 (a) $x^2 + 2x$ (b) $x^2 + 6x + 8$ (c) $x^2 + x - 6$
  (d) $x^2 + 2x - 15$ (e) $x^2 - 4x + 4$
9 (a) $x^2 + ax + 4x + 3a + 3$
  (b) $x^2 + ax + 2a - 4$
  (c) $x^2 + ax + 2x - a - 3$

Exercise 15B

3 (a) $6x^2 + 22a + 20$ (b) $9x^2 - 12x + 4$
  (c) $6x^2 - 8x - 8$ (d) $12a^2 - 23a + 10$

4 (a) $4x^2 + 12x + 9$ (b) $3x^2 - 14x + 8$
  (c) $4x^2 + 23x - 35$ (d) $16x^2 - 40x + 25$
5 (a) $6m^2 - mn - 12n^2$
  (b) $15a^2 - 34ab - 16b^2$
  (c) $12a^2 - 13ab - 14b^2$
  (d) $9a^2 - 12ab + 4b^2$
6 (a) 11 025 (b) 6561 (c) 7056
  (d) 11 881 (e) 6084

Chapter 16

Test yourself

1 and **2** See Figure A16:1.
3 See Figure A16:2.

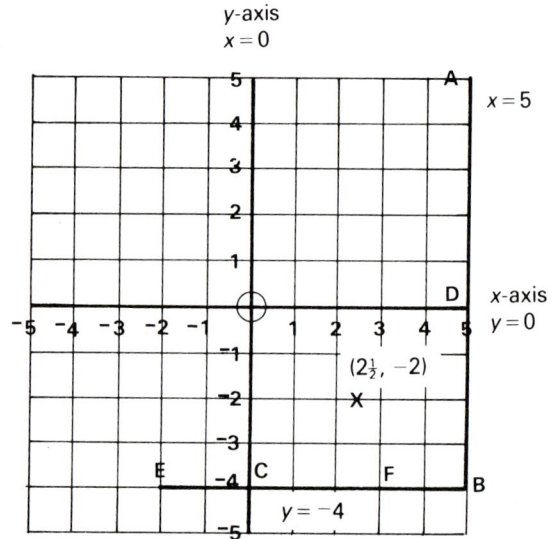

Fig. A16:1

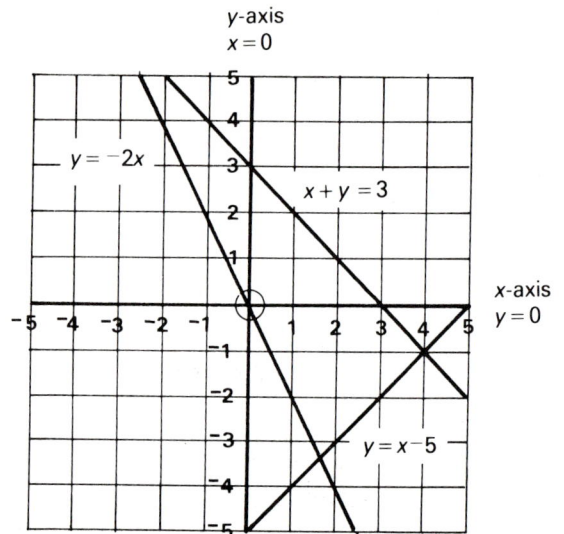

Fig. A16:2

4 $y = \frac{2}{3}x + 1$.

5 See Figure A16:3.

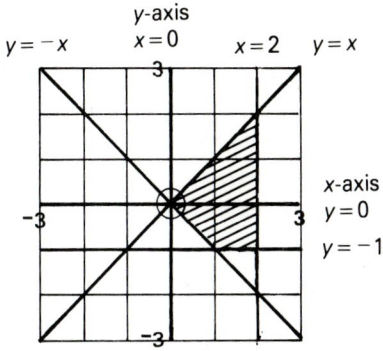

Fig. A16:3

Exercise

8 (a) (i) (0, 2) (ii) (0, −5) (iii) (0, a)
   (b) $x = 0$

9 See Figure A16:4.

(a) $y = x^2 + 1$

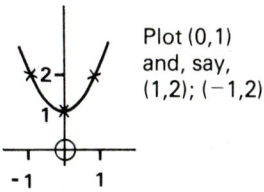

Plot (0,1)
and, say,
(1,2); (−1,2)

(b) $y = x^2 − 3$

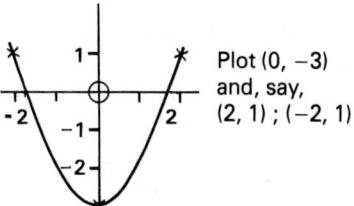

Plot (0, −3)
and, say,
(2, 1) ; (−2, 1)

Fig. A16:4

10 See Figure A16:5.

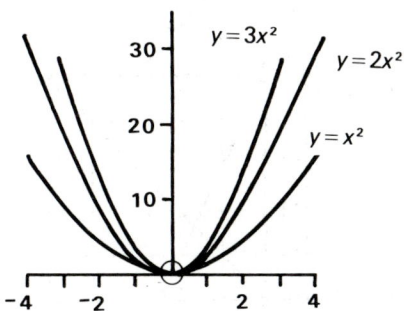

Fig. A16:5

11 See Figure A16:6.

14 See Figure A16:7.

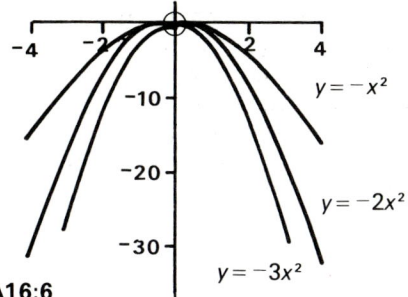

Fig. A16:6

(a)

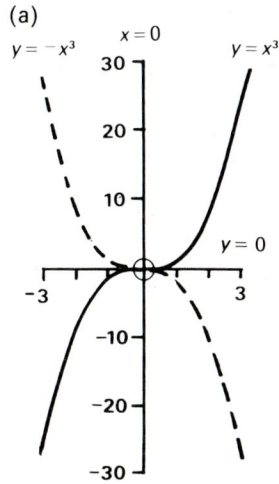

(b) $x = 0$

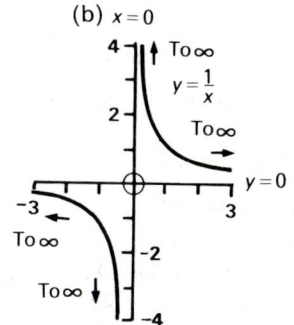

(c)

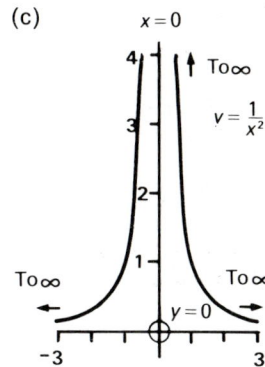

(d)

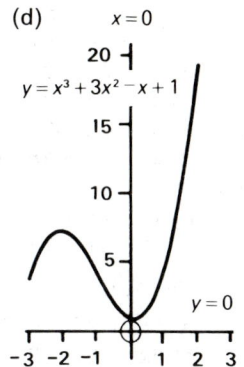

Fig. A16:7

Chapter 17

Exercise 17A

5 (a) $3x(1-2y)$ (b) $a(m-n)$
(c) $4ab(4c-3)$ (d) $3ab(x-2y)$
6 (a) $x(2x-1)$ (b) $3a(a+5)$
(c) $2xy(x-y)$ (d) $6mn(2m-3n^2)$
7 (a) $a(b-c)$ (b) $b(ac-3d)$
(c) $4y(3y-2)$ (d) no factors
8 (a) $2mp(2n+3)$ (b) $3mn(2n-3m)$
(c) no factors (d) $2\pi r(r+h)$
9 (a) no factors (b) $8x(3x-2)$
(c) $7h(3h-2n)$ (d) $ab(a-b^2)$
10 (a) $4a(1-4ab)$ (b) $3ab(1-a^2b)$
(c) $mx(ay-bd)$
11 (a) $2(2m-6n-3b+4d)$
(b) $3(2a+3b-4c-1)$
12 (a) $(x+y)(3+a)$ (b) $(a+5)(m+3)$
(c) $(a+1)(b+4)$ (d) $(1-2c)(5b-2d)$
(e) $(x+2y)(c-d)$ (f) $(a+2b)(p+d)$
13 (a) $\dfrac{2}{y}$ (b) $\dfrac{a}{3b}$ (c) $\dfrac{4a}{5b}$ (d) $\dfrac{2a}{3b}$

Exercise 17B

3 Check with a calculator!
4 (a) 2200 units^2 (b) 6600 units^2
(c) 2816 units^2
5 (a) $2(a+4)(a-4)$ (b) $2(2+x)(2-x)$
(c) $3(3+b)(3-b)$ (d) $3(5+b)(5-b)$
6 (a) $a(a+3)(a-3)$ (b) $2a(3+b)(3-b)$
(c) $2\pi r(r+h)$ (d) $(5x+6y)(5x-6y)$
(e) $(7x+1)(5x-1)$ (f) $(6+7x)(6-5x)$

Chapter 18

Exercise 18A

4 (a) d ($\angle$s in same segment)
(b) g ($\angle$s in same segment)
(c) f ($\angle$s in same segment)
(d) h $\angle$s in same segment)
(e) k (vert. opp. $\angle$s) (f) l (vert. opp. $\angle$s)
5 (a) 90° ($\angle$ in semicircle) (b) 30° ($\angle$ sum of Δ)
(c) 30° ($\angle$s in same segment)
6 (a) equal $\angle$s in isos. Δ
(b) $\angle$s in same segment (c) both $= d$
(d) a and c are equal alternate $\angle$s
7 (a) a diameter (b) $\angle$BAC: $\angle$BDC
8 $\angle$ACD = 50° ($\angle$s in same segment);
$\angle$DAC = 50° ($\angle$ sum of Δ)
9 (a) 90° (b) 90° (c) 90° (d) 90°
(e) both $+ c = 90°$ (f) $\angle$s in same segment
(g) both $= a$
10 $\angle$ABX = $\angle$AXB = $\angle$DXC and
$\angle$ABD = $\angle$ACD

Exercise 18B

6 ABCE; ABDE; BCDE; BDFE; CDFE
7 Square or rectangle.
8 (a) $g = 80°$; $e = 100°$ (b) $g = 60°$; $e = 120°$
(c) $g = 105°$; $e = 75°$
9 $\angle$ABC = 70°; $\angle$ADC = 110°;
∴ opposite $\angle$s total 180°
10 $\angle$AOX = 2$\angle$AYX;
$\angle$AYX = $\angle$CBX (both $+ \angle$ABX = 180°)
11 (a) $a = 80°$ (exterior $\angle$ equals opposite
interior $\angle$)
$b = 100°$ (opposite $\angle$s of cyclic quad.)
(b) $\angle$A $+ \angle$B = 180° (interior $\angle$s)
(c) $f = x$; $b = 180 - x$;
∴ $\angle$A $+ \angle$B = $x + 180 - x$ = 180°
13 Opposite $\angle$s total 180° (60° + 120°; 80° + 100°)
23 $\angle$C = 90°; $\angle$CBA = 60°; $\angle$CBD = 30°;
$\angle$CAD = 30°
14 $\angle$PQR = $\angle$PST; $\angle$PQR = $\angle$PRQ = $\angle$PSQ

Chapter 19

Exercise

9 (a) 1 (b) 1 (c) 1 (d) 4 (e) 3
10 (a) $9a^2$ (b) $-a^3$ (c) $27a^3c^3$ (d) -64
(e) $-a^3b^3$ (f) $81a^8$ (g) $4a^4b^2$ (h) $\dfrac{1}{b^6}$
(i) $\dfrac{8a^3}{27b^3}$ (j) $\dfrac{a^4}{b^6}$
11 (a) 3 (b) 5 (c) 3 (d) 5 (e) 2
12 (a) 9 (b) 8 (c) 8 (d) 27 (e) 32
(f) 125 (g) 32
13 (a) $\frac{1}{4}$ (b) $\frac{1}{3}$ (c) $\frac{1}{8}$ (d) 1 (e) $\frac{1}{4}$
14 (a) $\frac{1}{2}$ (b) $\frac{1}{2}$ (c) 6 (d) $\frac{1}{6}$ (e) 5
15 (a) $\frac{1}{9}$ (b) 8 (c) $\frac{1}{8}$ (d) $\frac{1}{27}$ (e) 32
16 (a) $\frac{1}{8}$ (b) $\frac{1}{64}$ (c) $\frac{1}{25}$ (d) $\frac{1}{1\,000\,000}$ (e) 16
17 (a) $\dfrac{1}{9a^2}$ (b) $\dfrac{3}{a^2}$ (c) $\dfrac{1}{2a}$ (d) $4a^2$ (e) $\dfrac{1}{16a^4}$
(f) 8 (g) $6a$ (h) 4 (i) $\frac{1}{3}$ (j) $6a$ (k) 3
(l) $1\frac{1}{4}$ (m) $2\frac{1}{2}$ (n) 1 (o) 4 (p) 0.2 (q) 1
(r) 0.09 (s) 1 (t) $\dfrac{1}{a^2}$

Chapter 20

Test yourself

1 (a) (i) 0.2 (ii) 0.35 (iii) 0.125
(b) (i) $\frac{1}{5}$ (ii) $\frac{7}{20}$ (iii) $\frac{1}{8}$
2 (a) 25% (b) 80% (c) $62\frac{1}{2}$% (d) $33\frac{1}{3}$%
(e) 42% (f) 2% (g) $62\frac{1}{2}$%
3 (a) £0.80 (b) £7.50 (c) 45 g (d) 21 m
(e) £1.30 (f) £3
4 (a) 25% (b) 20% (c) 80% (d) 300%
(e) 125%

5   (a) 60   (b) 18   (c) $62\frac{1}{2}$ kg   (d) 81.6 m
6   (a) 20% profit   (b) 10% loss
7   80%
8   375 g nitre, 50 g sulphur, 75 g charcoal
9   25%
10   £21
11   96%
12   13.5
13   £307.80
14   7 g
15   11.25 kg
16   $11\frac{1}{9}$%
17   12%
18   (a) $86\frac{2}{3}$%   (b) $46\frac{3}{7}$%
19   (a) £5300; £5830; £6063.20; £6669.52
      (b) 33%

Exercise

3   £5.44
4   50p
5   Rachel, by $13\frac{1}{3}$%.
6   £900
7   £60
8   £980
9   £363.75
10   £450
11   150 mm
12   $83\frac{1}{3}$ litres
13   £576
14   £2.65
15   (a) £3.20   (b) 25p

Chapter 21

Exercise 21A

3   (a)

−3	−2	−1	0	1	2
9	4	1	0	1	4
−6	−4	−2	0	2	4
1	1	1	1	1	1
4	1	0	1	4	9

(b)

−1	0	1	2	3	4
−1	0	1	4	9	16
2	0	−2	−4	−6	−8
2	2	2	2	2	2
5	2	1	2	5	10

(c)

0	$\frac{1}{2}$	1	$1\frac{1}{2}$	2	$2\frac{1}{2}$	3
0	$\frac{1}{4}$	1	$2\frac{1}{4}$	4	$6\frac{1}{4}$	9
0	$-1\frac{1}{2}$	−3	$-4\frac{1}{2}$	−6	$-7\frac{1}{2}$	−9
1	1	1	1	1	1	1
1	$-\frac{1}{4}$	−1	$-1\frac{1}{4}$	−1	$-\frac{1}{4}$	1

(d)

−1	0	1	2
2	0	2	8
−1	0	1	2
1	1	1	1
2	1	4	11

(e)

−1	0	1	2	3
1	1	1	1	1
−2	0	2	4	6
−1	0	−1	−4	−9
−2	1	2	1	−2

4   See Figure A21:1 (below and overleaf).
5   (a) $x = -1$ (b) $x = 1$ (c) $x = 1\frac{1}{2}$ (d) $x = -\frac{1}{4}$
      (e) $x = 1$
6   (c) $x = 0.25$ (d) $-10\frac{1}{8}$ when $x = 0.25$
7   (a) See Figure A21:1a.
      (b) −2 and 0   (d) −3 and 1
      (e) About −3.4 and 1.4; $x^2 + 2x = 5$
      (f) About −3.8 and 1.8; $x^2 + 2x = 7$
      (g) It is never true for any value of x.
8   (a) See Figure A21:1b
      (b) (i) $x = 1$; $x^2 - 2x = -1$
         (ii) $x = 0$ and $x = 2$; $x^2 - 2x = 0$
         (iii) $x = -1$ and $x = 3$; $x^2 - x = 3$
         (iv) About −0.7 and 2.7; $x^2 - 2x = 2$
         (v) About −1.6 and 3.6; $x^2 - 2x = 6$
9   (a) $x^2 - 3x = -1$ (b) $2x^2 + 3x = -1$
      (c) $x^2 - x = 3$ (d) $3x^2 - 2x = 1$
      (e) $2x^2 - 3x = 0$
11   (a) $y = x^2$ (b) $y = \frac{1}{2}x^2$ (c) $y = 2x^2$
      (d) $y = \frac{1}{4}x^2$ (e) $y = \frac{1}{8}x^2$

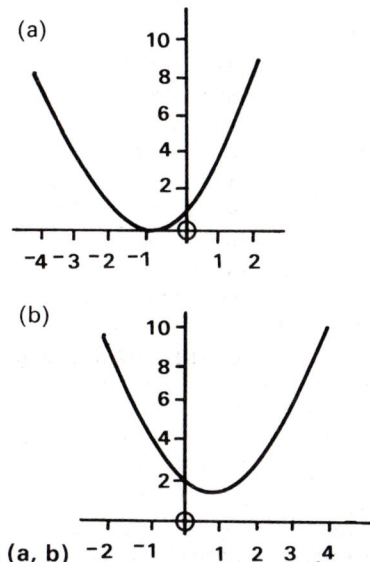

(a)

(b)

Fig. A21:1 (a, b)

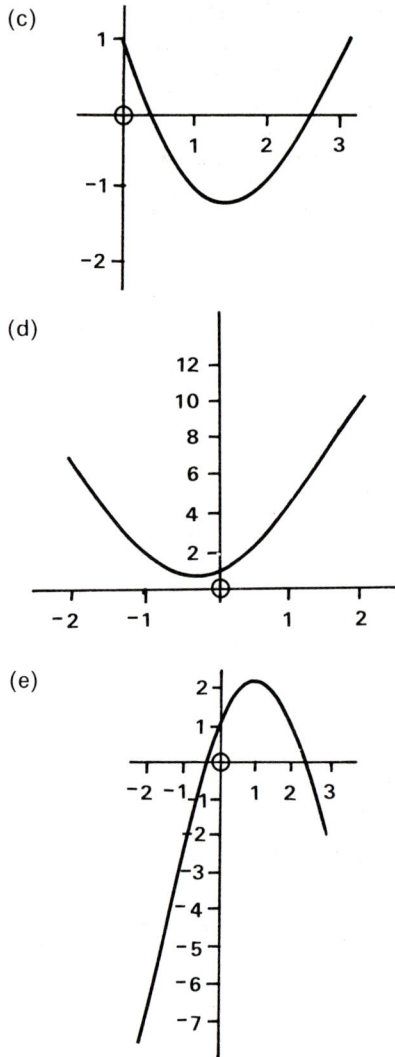

(c)

(d)

(e)

Fig. A21:1 (c–e)

Exercise 21B

1 $x^2 - 3x + 1 = 1$; $y = 1$; $x = 0$; $x = 3$
2 (a) $y = -1$ (b) $y = -6$ (c) $y = -4$
   (d) $y = -8$ (e) $y = 0$ (f) $y = -5$
3 (a) -4.45; 0.45 (b) -3.87; -0.13
   (c) -4.12; 0.12 (d) -3.58; -0.42
   (e) -4.55; 0.55 (f) -4; 0
4 $x^2 = -3x + 4$ or $x^2 + 3x - 4 = 0$;
   $x = -4$; $x = 1$
5 (a) $x^2 - 2x - 1 = 0$ (b) $x^2 + 2x - 1 = 0$
   (c) $x^2 - 3x + 2 = 0$ (d) $x^2 + 2x + 4 = 0$
   (e) $2x^2 - x - 6 = 0$ (f) $3x^2 - x + 3 = 0$
6 (a) -0.4; 2.4 (b) -2.4; 0.4 (c) 1; 2
   (d) no solution (e) -1.5; 2 (f) no solution

Chapter 23

Test yourself

1 (i) 4, 4, 4 (ii) 34, 34, 34 (iii) 18, none, 15.5
   (iv) 117, 109 and 116, 116 (v) 0, -4, -2
   (vi) 6, 8, 7.5
2 (a) 13 yr 4 mth (b) 10 min 20 s (c) 13 yr
3 24 °C
4 (a) (i) £110 (ii) £112
   (b) (i) £135 (ii) £112
5 £567
6 24 yr 2 mth
7 (a) 85 min (b) 14 min 10 s
8 (a) £262 (b) £26 (c) £26
9 (a) 3 °C (b) 5 °C (c) 5 °C
10 12 yr 7 mth
11 (a) 7 (b) 6.5 (c) 6.1
12 (a) 12 (b) 15 (c) 13
13 Tally: 0, 7, 21, 48, 24. Mean: 33.5
14 (a) 181 (b) £382
15 (a) 59.5 mm (b) 39 mm
16 69.4%
17 26 yr 11 mth
18 (a) (i) $5\frac{1}{7}$, 3, $4\frac{5}{7}$, $4\frac{6}{7}$ (ii) $4\frac{3}{7}$
    (b) Week 1, 0.0143%

Exercise

1   4.5   22.5
    14.5   87
    24.5   196
    34.5   345
    44.5   311.5
    55   220
    Mean $\simeq$ 30
2   0–9   //   2
    10–19 ЖHT ЖHT   10
    20–29 ЖHT ЖHT ЖHT //   17
    30–39 ЖHT ЖHT ЖHT ЖHT ЖHT ЖHT // 32
    40–49 ЖHT ЖHT ///   13
    50–59 ЖHT /   6
    30–39 is the modal class.
3   (i) 11.3 (ii) 13.75 (iii) 48.9 (iv) 11.9
    (Note: It would be sensible to round all these to
    the nearest unit. Why?)
4   (a) 56
    (b) (i) 11 (ii) 14 (iii) 47 (iv) 11
5   (a) 32.25 (b) 31 (c) 32.3 (d) 33

Chapter 24

Exercise

5   (a) 7.92 cm (b) 24.0° (c) 9.31 cm
    (d) 2.21 m (e) 44.4° (f) 18.5 mm

6 (a) AC = 8.87 cm; $\angle$C = 26.8°
    (b) HI = 4.92 cm; $\angle$I = 66.0°
    (c) KL = 5.86 cm; $\angle$K = 57.8°
    (d) MN = 3.45 m; $\angle$O = 57.4°
    (e) QR = 4.90 cm; $\angle$R = 45.6°
    (f) XV = 19.4 mm; $\angle$X = 72°

7 (a) $\angle$A = 70.8°; $\angle$B = 19.2°; AB = 9.16 cm
    (b) $\angle$F = 51.8°; FH = 11.7 m; GH = 14.9 m
    (c) $\angle$K = 35.2°; $\angle$J = 54.8°; KL = 21.4 cm
    (d) $\angle$D = 29.0°; DF = 32.4 cm; DE = 37.0 cm
    (e) Angles: 90°, 60°, 30°; QR = 5.3 km;
        PT = QP = 3.06 km;
        RS = PR = PS = 6.12 km
    (f) $\angle$Z = 54.8°; $\angle$X = 70.4°; XY = XZ = 15.3 m
    (g) $\angle$Q = $\angle$R = 68.3°; $\angle$P = 43.5°;
        PQ = 17.08 cm

8 (a) (i) 6.17 cm (ii) 85° (iii) 11.3 cm
    (b) (i) 30.4° (ii) 85.6° (iii) 9.26 cm
    (c) (i) 20.5 cm (ii) 65° (iii) 19.3 cm

9 (a) $\angle$A = 71.2°; $\angle$C = 68.6°; AB = 18.2 m
    (b) $\angle$D = 70.5°; DF = 8.74 cm; DE = 10.8 cm
    (c) $\angle$H = 27.9°; $\angle$I = 73.4°; GH = 8.84 cm

10 (a) $\frac{4}{5}$ (b) $\frac{5}{12}$

11 (a) 80° (b) 63.2° (c) 27.8°

12 (a) 104.1° (b) 12.0 cm (c) 7.45 cm

Chapter 26

Exercise

1 (a) 24 cm^2 (b) 12 cm^2 (c) 35 cm^2
    (d) 616 cm^2

2 (a) 420 cm^3 (b) 452 cm^3 (c) 40 cm^3
    (d) 12 500 cm^3 (e) 904 cm^3

3 (a) 40 cm^2 (b) 37.5 cm^2 (c) 50.2 cm^2

4 (a) 216 cm^3 (b) 312 cm^3 (c) 3170 cm^3
    (d) 147 cm^3 (e) 116 cm^3 (f) 3050 cm^3

5 (a) 462 000 cm^3 (b) 462 litres

6 (a) 72 cm^3 (b) 2.16 cm^3 (c) 0.707 cm^3
    (d) 0.53 cm^3

7 55 m^3

9 2640 cm^2; 10 400 cm^3

10 1430 cm^2; 3470 cm^3

11 100

12 80 m

13 99 h

14 (a) 78.5 cm^2 (b) 393 000 cm^3 (c) 23.6 m^3

15 (a) 29 600 litres (b) 1090 litres
    (c) 29 400 litres

16 1188

Chapter 27

Exercise

2 (c) 33 days (d) 40 days, 24 days, 16 days

3 (b) 14 scored 20 or less;
       17 scored more than 20.
    (c) 26 scored 28 or less;
       5 scored more than 28.
    (d) 2 scored 8 or less;
       29 scored more than 8.

4 (a) (i) 14 (ii) 17 (iii) 26 (iv) 5
    (b) (i) 24 (ii) 18 (iii) 14

5 (a) 14 (b) 19

6 (a) (i) 25 (ii) 12 (iii) 41 (iv) 29
    (b) (i) 35 (ii) 17 (iii) 56 (iv) 39

7 (a) 3.78 (b) 4.26 (c) 27.4 (d) 58.6

Chapter 28

Exercise

5 (b) 5.83

6 (a) $-\underset{\sim}{b}$ (b) $-\underset{\sim}{a}$ (c) $\underset{\sim}{b}$ (d) $-\underset{\sim}{b}$ (e) $\underset{\sim}{a}$
    (f) $-\underset{\sim}{a}$ (g) $\underset{\sim}{a} + \underset{\sim}{b}$

7 $\overrightarrow{SQ} = -\underset{\sim}{a} + \underset{\sim}{b}$; $\overrightarrow{QS} = -\underset{\sim}{b} + \underset{\sim}{a}$

8 (a) $-\underset{\sim}{x}$ (b) $-\underset{\sim}{y}$ (c) $2\underset{\sim}{z}$ (d) $-\underset{\sim}{z}$ (e) $-\underset{\sim}{z}$
    (f) $\underset{\sim}{x} + \underset{\sim}{z}$ or $\underset{\sim}{y} - \underset{\sim}{z}$ (g) $-\underset{\sim}{z} - \underset{\sim}{x}$ or $\underset{\sim}{z} - \underset{\sim}{y}$

9 (a) $-\underset{\sim}{b}$ (b) $\underset{\sim}{a}$ (c) $\frac{1}{2}\underset{\sim}{b}$ (d) $-\underset{\sim}{a} + \frac{1}{2}\underset{\sim}{b}$
    (e) $\frac{1}{2}\underset{\sim}{b} + \underset{\sim}{a}$

10 (a) $2\underset{\sim}{c}$ (b) $-\underset{\sim}{d}$ (c) $\underset{\sim}{d}$ (d) $2\underset{\sim}{d}$ (e) $\underset{\sim}{c} + \underset{\sim}{d}$
    (f) $\underset{\sim}{c} + 2\underset{\sim}{d}$ (g) $-\underset{\sim}{c} + 2\underset{\sim}{d}$ (h) $-\underset{\sim}{c} + \underset{\sim}{d}$

11 (a) $\overrightarrow{AC}$ (b) $\overrightarrow{BF}$

13 See Figure A28:1

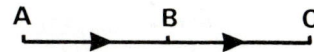

Fig. A28:1

14 (a) $3\underset{\sim}{x}$ (b) $2\underset{\sim}{x}$

15 (a) If $\underset{\sim}{c} = \underset{\sim}{d}$ then they would have to be parallel.
       (This is very important. Remember it!)
    (b) As $\underset{\sim}{c} \neq \underset{\sim}{d}$ then this can only be true if
       $h + k = 0$ and $h - k + 1 = 0$, $\therefore h = -\frac{1}{2}$ and
       $k = \frac{1}{2}$.

16 $\overrightarrow{BC} = -\overrightarrow{AB} + \overrightarrow{AC}$; $\overrightarrow{BM} = \frac{1}{2}(-\overrightarrow{AB} + \overrightarrow{AC})$;
    $\overrightarrow{AM} = \overrightarrow{AB} + \frac{1}{2}(-\overrightarrow{AB} + \overrightarrow{AC}) = \frac{1}{2}\overrightarrow{AB} + \frac{1}{2}\overrightarrow{AC} = \frac{1}{2}(\overrightarrow{AB} + \overrightarrow{AC})$

17 (a) Parallel, with XY : BC = h : 1 (or 'XY is h
       times as long as BC')
    (b) Hint: Show that $\overrightarrow{XY} = h(-\overrightarrow{AB} + \overrightarrow{AC})$.

Chapter 29

Exercise

11 (a) $(3x - 4)(2x - 1)$ (b) $(3x + 4)(2x + 1)$
    (c) $(3x - 4)(2x + 1)$ (d) $(3x + 4)(2x - 1)$

12 (a) $(x - 3)(x + 2)$ (b) $(x - 3)(x - 2)$
  (c) not possible   (d) not possible
  (e) not possible   (f) $(3x + 2)(x - 1)$
  (g) $(3x + 2)(x + 3)$ (h) not possible
  (i) not possible   (j) not possible
  (k) $(4x + 1)(x + 1)$ (l) $(2x + 3)(3x + 2)$
  (m) $(3x + 4)(x + 3)$ (n) $(6x + 1)(x - 4)$
  (o) $(4x + 1)(2x + 3)$ (p) $(4x + 5)(2x - 3)$
13 (a) $3(x + 2)(x + 2)$ (b) $2(x + 3)(x + 4)$
  (c) $3(x - 1)(x - 4)$ (d) $3(3x - 1)(x + 1)$
  (e) $2(2x - 1)(x - 2)$ (f) $4(3x - 8)(x + 2)$
14 (a) $(3a - b)(3a - b)$ (b) $(7 - x)(5 - x)$
  (c) $(7 + x)(5 - x)$ (d) $(3x - y)(x + 2y)$
  (e) $(4x - 3)(2x - 3)$ (f) $(5x - 3y)(5x - 3y)$
  (g) $(x + \frac{1}{2})(x + \frac{1}{2})$ (h) $(x - \frac{1}{2})(x + 1\frac{1}{2})$
  (i) $(6 + x)(4 - x)$

Chapter 30

Exercise 30A

7 (a) £127.50   (b) £662.50   (c) 8%   (d) 7%
  (e) £2500   (f) £1500   (g) 8%   (h) £25
  (i) £2500   (j) £35.10   (k) $3\frac{1}{2}$%   (l) £12 500
8 £7500
9 £315
10 (a) £4470   (b) £370   (c) June, £30

Exercise 30B

3 (a) £655.40   (b) £125.88   (c) £31.06
  (d) £14 909.02
4 £25, £26.25, £27.56, £28.94, £30.39, £31.91
5 £9500, £9025, £8573.75, £8145.06, £7737.81
6 (a) £701.27   (b) £1607.69   (c) £1948.40
  (d) £1518.07   (e) £62 458.50
7 (a) £822 973   (b) £90 900   (c) £63 000

Chapter 31

Exercise 31A

11 (a) 5, −5   (b) −4, 8   (c) −9, 10
  (d) −11, 2   (e) 1, 6   (f) 0, $\frac{3}{7}$
12 (a) 2 m by 10 m   (b) 11 cm side   (c) −13 or 3
  (d) $-\frac{1}{5}$ or $\frac{9}{2}$
13 (a) $x^2 - 6x + 8 = 0$ (b) $x^2 - 4 = 0$
  (c) $x^2 - 2x - 15 = 0$ (d) $x^2 + 6x + 8 = 0$
  (e) $x^2 + 4x = 0$ (f) $6x^2 + x - 2 = 0$
14 (a) $(3, 0)$; $(−1, 0)$; $(0, −3)$
  (b) $(4, 0)$; $(−1, 0)$; $(0, −4)$
  (c) $(3, 0)$; $(−3, 0)$; $(0, −9)$
  (d) $(−2, 0)$ (tangential); $(0, 4)$
  (e) $(1\frac{1}{2}, 0)$; $(1, 0)$; $(0, 3)$
  (f) $(−3, 0)$ (tangential); $(0, 9)$

15 See Figure A31:1

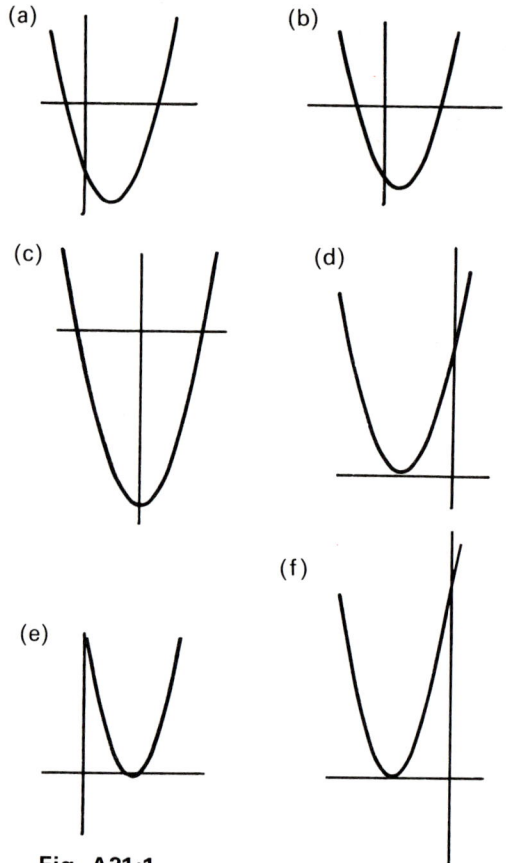

Fig. A31:1

16 (a) 12 cm   (b) 7   (c) 20

Exercise 31B

1 (a) −4.24, 0.24   (b) −1.45, 3.45
  (c) −4.30, −0.70
2 (a) −1.42, 8.42   (b) −6.65, −1.35
  (c) −2.30, 1.30
3 (a) −1.37, 0.37   (b) 0.29, 1.71
  (c) −2.14, 0.47
4 (a) −0.79, 3.79   (b) −1.71, −0.29
  (c) −0.77, 0.43
5 (a) $2x^2 - 4x - 1 = 0$ (b) $3x^2 - 4x + 1 = 0$
  (c) $2x^2 - x - 2 = 0$ (d) $2x^2 + 4x - 3 = 0$
  (e) $x^2 - 3x + 3 = 0$ (f) $x^2 - 4x + 4 = 0$
6 (a) −4.12, 0.12   (b) −5.85, 0.85
  (c) −0.54, 1.87
7 (a) $3x^2 + 9x + 7 = 0$ (b) $8x^2 - 3x + 2 = 0$
  (c) $3x^2 - 2x + 5 = 0$ (d) $3x^2 - 2x - 1 = 0$
  (e) $9x^2 - 6x + 1 = 0$ (f) $4x^2 - 5x - 2 = 0$
8 (c) A: 5(a), 5(b), 5(c), 5(d), 7(d), 7(f)
  B: 5(e), 7(a), 7(b), 7(c)
  C: 5(f), 7(e)

Chapter 32

Exercise 32B

5 50.9 cm^3

6 340 cm^3

7 $\frac{1}{8}$

8 7.5 × 10^5 litres

9 (c) (i) 8.94 cm   (ii) 9.79 cm   (iii) 26.6°
   (iv) 24.1°

10 (b) 12.1 cm   (b) 19.3°

11 6.93 cm

12 Yes (longest diagonal = 3.74 cm).

13 (b) 10.8 cm   (c) 19.7 cm   (d) 20.1 cm
   (e) 21.8°; 11.5°

14 (c) ∠BCF

15 (a) 36.9°   (b) 15.6°

16 5.77 cm

17 43 cm^3

Index

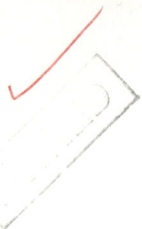